AF849013

URBAN DEVELOPMENT AND INFRASTRUCTURE

URBAN AND BUILT ENVIRONMENTS

SUSTAINABLE DEVELOPMENT, HEALTH IMPLICATIONS AND CHALLENGES

URBAN DEVELOPMENT AND INFRASTRUCTURE

Additional books in this series can be found on Nova's website under the Series tab.

Additional e-books in this series can be found on Nova's website under the e-book tab.

URBAN DEVELOPMENT AND INFRASTRUCTURE

URBAN AND BUILT ENVIRONMENTS

SUSTAINABLE DEVELOPMENT, HEALTH IMPLICATIONS AND CHALLENGES

ALEXIS COHEN
EDITOR

nova publishers
New York

Copyright © 2015 by Nova Science Publishers, Inc.

All rights reserved. No part of this book may be reproduced, stored in a retrieval system or transmitted in any form or by any means: electronic, electrostatic, magnetic, tape, mechanical photocopying, recording or otherwise without the written permission of the Publisher.

We have partnered with Copyright Clearance Center to make it easy for you to obtain permissions to reuse content from this publication. Simply navigate to this publication's page on Nova's website and locate the "Get Permission" button below the title description. This button is linked directly to the title's permission page on copyright.com. Alternatively, you can visit copyright.com and search by title, ISBN, or ISSN.

For further questions about using the service on copyright.com, please contact:
Copyright Clearance Center
Phone: +1-(978) 750-8400 Fax: +1-(978) 750-4470 E-mail: info@copyright.com.

NOTICE TO THE READER

The Publisher has taken reasonable care in the preparation of this book, but makes no expressed or implied warranty of any kind and assumes no responsibility for any errors or omissions. No liability is assumed for incidental or consequential damages in connection with or arising out of information contained in this book. The Publisher shall not be liable for any special, consequential, or exemplary damages resulting, in whole or in part, from the readers' use of, or reliance upon, this material. Any parts of this book based on government reports are so indicated and copyright is claimed for those parts to the extent applicable to compilations of such works.

Independent verification should be sought for any data, advice or recommendations contained in this book. In addition, no responsibility is assumed by the publisher for any injury and/or damage to persons or property arising from any methods, products, instructions, ideas or otherwise contained in this publication.

This publication is designed to provide accurate and authoritative information with regard to the subject matter covered herein. It is sold with the clear understanding that the Publisher is not engaged in rendering legal or any other professional services. If legal or any other expert assistance is required, the services of a competent person should be sought. FROM A DECLARATION OF PARTICIPANTS JOINTLY ADOPTED BY A COMMITTEE OF THE AMERICAN BAR ASSOCIATION AND A COMMITTEE OF PUBLISHERS.

Additional color graphics may be available in the e-book version of this book.

Library of Congress Cataloging-in-Publication Data

ISBN: 978-1-63483-067-6

Published by Nova Science Publishers, Inc. † New York

CONTENTS

PREFACE

The relationship between human beings and the urban environment is a topic increasing in relevance. Most of the people in the world live in an urban context and, as many scholars assert, the more urbanism is developing, the more specific problems arise, such as insecurity, loneliness and weak social ties. This book explores urban and built environments, focusing on chapters that examine the predictors of life satisfaction from living in an urban environment; contradictions between environmental indicators for a walkable neighbourhood and human behavior; and an evaluation of changes in urban environments using orthophoto maps (i.e., aerial images). This kind of research is very important for sustainable development of urban environments to gain control of the development of construction sites to monitor the keeping of building regulations and the use of building materials.

Other chapters explore volatile organic compounds (VOC) based on their detrimental effects on human health and the environment; key factors for the sustainability of built environments; and efficiency of planning and design tools towards eco-friendly, high-density communities.

Chapter 1 - The relationship between inhabitants and their place of residence is important for people quality of life and well-being. This relationship entails both empirical and political challenges. Empirical challenges pertain to the investigation of different dimensions affecting well-being, distinguishing among personal characteristics (i.e., gender, age, income, status, education, etc.), psychosocial dimensions (i.e., place attachment, sense of community, place identity, etc.) and contextual aspects (objective dimension such as place dimension, services development, etc.). Political challenges refer to the importance of connecting different levels of intervention, in line with the World Health Organization approach "Health in all Policies."

On these grounds the study here presented aimed at investigating the best predictors of life satisfaction, considering both personal characteristics (i.e., gender, age, and income) and contextual psychosocial variables (i.e., place attachment and social identification with inhabitants). Participants were 251 community people (53.8% females) living in a big city in the North-West of Italy. Their mean age was 33.91 (SD = 15.79). Data were collected via a self-reported questionnaire, which took about 20 minutes to complete. Results showed that place attachment is the most powerful predictor of life satisfaction, followed by income and lower age, whereas social identification with inhabitants has no influence.

These findings are discussed from a psychosocial standpoint, considering also future challenges in public health involving urban environments.

Chapter 2 - The topic of walkable neighbourhoods is particularly prominent in the literature on Sustainable Cities in the Anglo-Saxon world. There is convincing evidence that segregated, spread-out land uses typical of urban sprawl (e.g., long distances between residential, business, industrial, shopping and recreational functions) have led to an amount of car reliance that impacts drastically on the opportunities, but also the willingness of urban residents to undertake their trips on foot. There is likely to be a health aspect, with numerous studies showing that people who live an active lifestyle and walk a lot are less exposed to a number of diseases and health-related effects that are usually attributed to an inactive, sedentary lifestyle (e.g., non-communicable diseases like obesity, coronary heart diseases and type-2 diabetes). Spatial indicators like the Walkability Index or the Walkscore have been developed in this context and help geographical and business analysts to highlight problem areas, or vice versa to stimulate demand advantages for walkable neighbourhoods on the real estate market.

Despite these efforts, the number and distance of walking trips in the Anglo-Saxon world remain amongst the highest in the world. Comparisons with mainland Europe show that urban residents walk and bike a lot more throughout all social classes, especially in parts of Europe where the climate is not always favourable of active lifestyles (The Netherlands, Scandinavia, Germany). In this context, recent adaptations of the Walkability Index and Walkscore for the City of Stuttgart, Germany, have shown that the urban structure approach they are based upon does not necessarily explain the amount that people walk very well. In a comparison with a large household survey database, people who walked less did not necessarily live in the least walkable areas and vice versa. The conclusion was that (1) there must be either other environmental forces at work that are not captured by the aforementioned indicators, or that (2) cultural and behavioural differences are predominant and environment does not play that much of a role after all.

This article systematically looks at a more comprehensive set of environmental conditions and personal characteristics in the neighbourhoods of survey respondents, with the aim to identify variables that play a specific role in the European context. It briefly reviews the literature on behavioural determinants for walking and gives an overview over current concepts to measure walkable neighbourhoods and highlights their shortcomings. In the methodological chapter, multi-level regression analysis is used to control the survey dataset for personal precursors for walking like age, employment, health status, and transport mode options. Based on these results, the article determines conceptual gaps in current walkability assessments and analyses what other parameters of urban form than the ones used by current walkability assessment tools have explanatory power, and what factors can be attributed to behavioural differences. The final chapter discusses what city planners and architects can learn from the results in order to deliver walkable neighbourhoods that stimulate walking for health more effectively.

Chapter 3 - Nowadays, urban environments are changing and expanding very fast, since people are moving to cities to get jobs, more opportunities and generally a better life. To gain control of this fast development and building projects, orthophotography images are used to display actual land and building features. In our research, the aerial images, acquired by a model multicopter equipped with a digital camera, were analyzed and compared to the images of orthophoto maps available on selected web providers such as Google Earth, Geopedia, Apple Maps and Bing Maps. Analyzed images were obtained/sampled from the selected area in Slovenia, in urban environment: The city of Ljubljana. Analysis was based on pedestrian areas, traffic areas, rooftops and green areas. In addition, acquired colors of Ljubljana (from the multicopter with a digital camera and orthophoto maps) were compared to the reference colors obtained from previous research.

The results of the analysis will show the differences between images acquired from the multicopter with a digital camera and images obtained from orthophoto maps. The research will also present advantages and disadvantages of using two different methods for obtaining images from the selected area of the city environment. Even more, comparison of colors that are typical for the selected city area will give the optimal method for the preservation of landscapes. This kind of research is very important for sustainable development of urban environments to gain control of the development of construction sites to monitor the keeping of the building regulations and the use of the building materials (i.e., according to the construction regulations and the requirements for landscape preservation). Since the weather patterns are changing very fast, the results of this research will also give important insight on the environmental changes.

Chapter 4 - The identification of VOC emission sources and analysis of their impact on air quality in complex Belgrade (Serbia) urban environment is presented. The concentrations of VOC were measured by using Proton Transfer Reaction Mass Spectrometer, whereas the concentrations of other pollutants (NO_x, NO, NO_2, SO_2, CO, and PM_{10}) and meteorological parameters were obtained from the automatic monitoring station located at the same site. Two US EPA receptor models, Positive Matrix Factorization and Unmix, were applied on the obtained dataset, and the resolved profiles were analyzed and compared. The divergent results suggesting weaknesses in used models are discussed in details. In addition to this, statistical analyses including bivariate polar plot, bivariate cluster and conditional bivariate probability function were performed within the statistical software environment R using the Openair package. The results enabled reliable identification of the emission sources with regards to several factors, such as: pronounced daily and weekly variations, day-of-week difference, statistical indicators, relationships with meteorological data - particularly wind speed and direction, toluene to benzene ratios, etc. Apart from the local sources attributed to vehicle exhaust, aged plumes, solid fuel burning and local laboratory, significant contributions of industrial facilities, including oil refinery, petrochemical complex, and chemical industry located in the suburban area of Belgrade were observed.

Chapter 5 - The accessibility and the location of new and existing urban projects are key factors for the sustainability of our built environments. Previous scientific research have shown that both the location of neighborhoods (that is to say their proximity to work places, schools facilities, services etc.) as well as their accessibility (proximity of infrastructures, bus services, etc.) have a huge impact on modal share (especially the ability to use active commuting for daily travel), travelled distances and thus on emissions of greenhouse gas and health of inhabitants. In a view of sustainability, the location of new and planned urban projects should be carefully assessed in order to favor the development of local communities encouraging the use of public transportation and active commuting (foot, bicycle). Local authorities should also be able to assess the potentialities of different Brownfield and Greenfields on their territories and define where to authorize the development of new urban projects, in order to inscribe them in a concrete vision of sustainability. In the first part of the chapter, we propose a synthetic review of the scientific literature dealing with the relationships between location, accessibility and commuting. Then, we analyze how location and accessibility are taken into account in several existing neighborhood sustainability assessment tools. Based on the ascertainments that sustainable mobility, location and accessibility criteria are given less importance than criteria related to building energy performance or green spaces and, that existing sustainability assessment tools are mainly developed for experts, we propose two frameworks to assess the location and the accessibility of buildings and neighborhoods. These two frameworks are especially designed for a non-expert audience (citizens, local authorities, developers, etc.), within the Walloon (Belgian) context. To conclude the chapter, the reproducibility of our work is discussed and our main findings are summarized.

Chapter 6 - High-density development in a city or community has been increasingly promoted in urban planning arena to achieve sustainable, ecological goals through impact on transportation behaviour and less per capita land consumption. Such high-density development to achieve region-wide ecological purpose, however, may have a negative ecological impact at the community level in terms of worsened natural light, wind, and the space needed for humans, animals, and vegetation to live healthily, and in turn affects the acceptance of high-density policy. One of the solutions to mitigating the negative impact may lie in the spatial arrangement of building mass and outdoor space since they determines the fundamental spatial characteristics in the community; however, few studies have been carried out in this area. This chapter aims to present an inventory of urban planning and design tools for arranging building mass and open space, and to examine their impacts on some eco-friendly indexes. A real case of urban renewal site is applied for simulation analysis to assess the impacts of the planning tools, while in order to evaluate the general impact of each planning tool a hypothetical city is developed, where other factors can be controlled for. The results suggest that lowering the building coverage rate and increasing height-distance ratio are the two most efficient tools for improving the overall quality of eco-cities for the selected indexes. The efficiency of the tools in improving individual aspects of an eco-city, such as pedestrian-friendly environment, green space accessibility, adaptive capacity of CO_2 absorption, and animal biodiversity are also gauged. These research findings can help improve the eco-city planning or design guidelines, which may work more effectively citywide in developing eco-city than the more labor-intensive case-based simulation review process, popular in many cities.

In: Urban and Built Environments
Editor: Alexis Cohen
ISBN: 978-1-63483-067-6
© 2015 Nova Science Publishers, Inc.

Chapter 1

LIVING IN THE URBAN ENVIRONMENT: A STUDY ON PREDICTORS OF LIFE SATISFACTION

Stefano Tartaglia, PhD,
Norma De Piccoli*, PhD and Chiara Rollero, PhD
Department of Psychology, University of Turin, Italy

ABSTRACT

The relationship between inhabitants and their place of residence is important for people quality of life and well-being. This relationship entails both empirical and political challenges. Empirical challenges pertain to the investigation of different dimensions affecting well-being, distinguishing among personal characteristics (i.e., gender, age, income, status, education, etc.), psychosocial dimensions (i.e., place attachment, sense of community, place identity, etc.) and contextual aspects (objective dimension such as place dimension, services development, etc.). Political challenges refer to the importance of connecting different levels of intervention, in line with the World Health Organization approach "Health in all Policies."

On these grounds the study here presented aimed at investigating the best predictors of life satisfaction, considering both personal

* Correspondence concerning this article should be addressed to: Norma De Piccoli, Dipartimento di Psicologia, Università di Torino, Via Verdi, 10, 10124 Torino, Italy. Phone: ++390116702024 ; Fax: ++390116702061; E-mail: norma.depiccoli@unito.it.

characteristics (i.e., gender, age, and income) and contextual psychosocial variables (i.e., place attachment and social identification with inhabitants). Participants were 251 community people (53.8% females) living in a big city in the North-West of Italy. Their mean age was 33.91 (*SD* = 15.79). Data were collected via a self-reported questionnaire, which took about 20 minutes to complete. Results showed that place attachment is the most powerful predictor of life satisfaction, followed by income and lower age, whereas social identification with inhabitants has no influence.

These findings are discussed from a psychosocial standpoint, considering also future challenges in public health involving urban environments.

INTRODUCTION

The relationship between human beings and the urban environment is a topic increasing in relevance. Most of people in the world live in an urban context and, as many scholars assert, the more urbanism is developing, the more specific problems arise, such as insecurity, loneliness, and weak social ties. Studying the consequences the urban context has on wellbeing and quality of life is an ethical, political, and scientific imperative. Addressing these effects is also a relevant political challenge.

Political challenges refer to the importance of connecting different levels of intervention. Social and community psychology demonstrated how different variables affect people's ordinary life, i.e., at individual (Gattino, Rollero, & De Piccoli, 2015), ideological (Rollero & Tartaglia, 2013; Rollero, Glick, & Tartaglia, 2014) and macro social level (Gattino & Tartaglia, 2015). Dealing with quality of life means to act in line with the World Health Organization approach "Health in all Policies." The Biopsychosocial Model (Engel, 1977) considers health as a combination of biological, psychological, and social factors following the World Health Organization (WHO) definition of health as a state of complete physical, mental, and social well-being, and not just as the absence of disease. A very large amount of research has studied quality of life considering the complexity of its determinants (e.g., Cummins, 2005; De Piccoli, 2014; Dzuka, 2012; Rollero, 2013a; Sirgy, 2011; Tartaglia, 2013; Tartaglia, 2014).

In psychological literature, there are a number of ways to conceptualize and measure quality of life and well-being of individuals (for a review see Lent, 2004; Meléndez, Tomas, Oliver, & Navarro, 2009). In Diener's (2006) perspective, subjective well-being refers to all the various type of evaluations, both positive and negative, that people make of their lives. It is an umbrella term for the different valuations people make regarding their lives, the events happening to them, their bodies and minds, and the circumstances in which they live. The cognitive component of subjective well-being has been defined as life satisfaction (Diener, 2006). The life satisfaction represents a broad, reflective appraisal the person makes of his or her life. The term life can be defined as all areas of a person's life at a particular point in time or as an integrative judgment about the person's life since birth, and this distinction is often left ambiguous in current measures (Diener, 2006). Overall, literature has focused on specific contextual factors that may affect life satisfaction and well-being, paying particular attention to the place of residence. The places affect human activities providing resources and setting limits, but also have psychological impact provoking relevant feelings and furnishing a base for cognitive elaboration.

The relationship between individuals and their environment has two linked but different aspects: an affective dimension, that is the emotional bond toward places (place attachment), and a cognitive dimension, related to the cognitions about the self as a member of a physical space (place identification) (Altman & Low, 1992; Hernández, Hidalgo, Salazar-Laplace, & Hess, 2007; Rollero, 2013b; Rollero & De Piccoli, 2010a). These concepts have been considered in relation to different territorial dimensions (i.e., the nation, the town, and the neighbourhood) (Lewicka, 2010). They involves different elements: the place, considering both the social relationship (bonding and bridging: Lewicka, 2011) and the physical characteristics; the personal characteristics of people living there; and the process involved, that is the affect, the cognition and the behaviour (Scannell, Gifford, 2010).

Specifically, place attachment can be defined as the affective link that people establish with specific environments, where they have a propensity to remain and where they feel comfortable and safe (Hidalgo & Hernández, 2001; Mannarini, Tartaglia, Fedi, & Greganti, 2006). It encourages greater freedom of behaviour, exploration, confidence and affective responsiveness within the local community (Fried, 2000). Moreover, the place can be considered a social category and thus subject to the same rules as a social identification. Therefore, place identification expresses membership of a group of people who are defined by their location (TwiggerRoss & Uzzell, 1996). Indeed, it applies specifically to the aspects of identity and those of the self-categorization on the basis of the membership to a locally defined group.

Literature shows that people's relation to their living environment is a key issue for their well-being (Rollero & De Piccoli 2010b; Rollero, Gattino, & De Piccoli, 2014), and for physical and mental health (Kawachi & Berkman 2003; Wen, Hawkley, & Cacioppo, 2006). This is especially true in the big cities and in general in the urban environment. Empirical studies shows that the feeling of being at home in the neighbourhood is closely connected with a sense of well-being (Moser & Ratiu, 2002), whereas the lack of environmental qualities may be a threat to quality of life (Uzzell & Moser, 2006). However, most studies have been focused only on the role of place attachment in predicting individuals' well-being. Since place identification expresses a significant aspect of identity connected to the local context, it should be important including this dimension in investigating the relationship between environmental bonds and well-being.

The Current Study

The present study aimed at investigating the best predictors of life satisfaction, considering both personal characteristics (i.e., gender, age, and income) and contextual psychosocial variables (i.e., place attachment and social identification with inhabitants). On the grounds of the above-described literature, we set two objectives:

1. Assessing the relative contribution of personal socio-demographic characteristics and psychosocial variables in explaining life satisfaction.
2. Comparing the effects of place attachment and social identification with inhabitants as predictors of life satisfaction.

Method

Data were collected via a self-reported questionnaire, which took about 20 minutes to complete. Participants were recruited among inhabitants of a middle-class neighborhood in Turin, a large city of about one million inhabitants in the North-West of Italy, via a convenience sampling. Although the sampling technique does present limitations in that it was not purely random, every attempt was made to access a wide range of respondents in terms of the age range considered, gender, educational level and occupational status. Anonymity of the respondents was assured.

Participants

The participants were 251 adults (53.8 % females). They aged from 19 to 74 (average age = 33.91; SD = 15.79). Most of them were unmarried (62%), 27.6% were married and 10.4% divorced or widowed. About the education, the majority was high school (60.8%) or college graduated (25.2%), whereas the others (14%) had a lower level of education. Half of the participants (50%) were workers, followed by students (41.9%), and retired or unemployed people (8.1%). Concerning the monthly income, 37% of participants declared a family income lower than 1200 Euros, 29.2% reported an income between 1200 and 2000 Euros, 20.1% between 2001 and 3000 Euros, and the remaining 13.7% more than 3000 Euros.

Measures

The variables used in our analysis were:

Life satisfaction. It was assessed using The Satisfaction with Life Scale (Diener, Emmons, Larsen, & Griffin, 1985) composed by 5 items (e.g., "In most ways my life is close to my ideal"; "If I could live my life over, I would change almost nothing") rated on a 7-point Likert-type scale ranging from 1 (Strongly disagree) to 7 (Strongly agree). The measure demonstrated good internal consistency (Cronbach's alpha =.90).

Place attachment. We used the Residential Attachment Scale (Bonaiuto, Fornara, Aiello, & Bonnes, 2002), a unifactorial Italian scale created to measure place attachment, made up of 8 items, referred to the neighbourhood (e.g., "By now my neighbourhood is a part of me"; "For me leaving my neighbourhood would be really hard"). Items were scored on a 7-point Likert-type scale ranging from 1 (Strongly disagree) to 7 (Strongly agree). The measure showed good internal consistency (Cronbach's alpha =.89).

Social identification with inhabitants. It was investigated using the Identification with a Psychological Group Scale (Mael & Tetrick, 1992; Rollero & De Piccoli, 2010a), an unifactorial scale used to assess the identification toward the social group of neighbours, made up of 10 items (e.g., "When someone criticizes my neighbours, I feel personally insulted";

"When I talk about my neighbours usually I say we and not they"). Items were rated on a 7-point Likert-type scale ranging from 1 (Strongly disagree) to 4 (Strongly agree). The measure demonstrated good internal consistency (Cronbach's alpha =.87).

Socio-demographic items. We collected the following data: age, gender, marital status, educational level, employment status and income.

Data Analysis

Besides descriptive statistics, to test the hypotheses we performed a multiple regression analysis. In this model life satisfaction was regressed onto demographic variables (gender, age, and income), place attachment and social identification with inhabitants. Statistical analyses were carried out using the SPSS 21.0 software.

Results

Table 1 shows descriptive statistics and correlations among life satisfaction, place attachment, and social identification with inhabitants. All the studied variables were positively correlated. The strongest relation was observed between place attachment and social identification with inhabitants.

Table 1. Descriptive statistics and correlations among the studied variables

	1	2	3
Life satisfaction (1)		.37**	.25**
Place attachment (2)			.52**
Social identification with inhabitants (3)			
Mean	4.28	3.64	2.20
SD	1.31	1.18	.64

** p<.01.

In the regression model the dependent variable was life satisfaction (Table 2). The model accounted for a significant amount of variance of Satisfaction with Life (adjusted $R^2 = .24$). Among socio-demographic variables, only gender played no significant role. Increasing age leads to less satisfaction, whereas income positively affected the dependent variable. Social identification with inhabitants did not result as a significant predictor. On the contrary, place attachment not only increased life satisfaction, but also was the most effective of the independent variables.

Table 2. Multiple regression analysis predicting life satisfaction

Predictor	β	t
Gender (0=male, 1=female)	.00	.05
Age	-.15*	-2.33
Income	.26***	4.27
Place attachment	.34***	4.89
Social identification with inhabitants	.12	1.64

*p<.05 ***p<.001 R^2 Adj. = .24 F (5, 216)= 14.36 p<.001.

DISCUSSION

The present study aimed at investigating the predictors of life satisfaction for people living in the urban environment. The place of residence, specifically the neighbourhood for the inhabitants of big cities, is still, in the globalised world, a relevant point of reference (Tartaglia, 2009; Tartaglia & Rossi, 2015). Several studies showed how the relation with place of residence may affect the people's quality of life (Kawachi & Berkman 2003; Rollero & De Piccoli 2010b; Wen et al. 2006). For these reasons, we have investigated as predictors of life satisfaction both personal characteristics classically associated to high-level life conditions, and contextual psychosocial variables assessing the quality of the relation with the neighbourhood of residence. As expected, age and income were significant predictors of life satisfaction. The increasing age reduced life satisfaction whereas income increased it. Nevertheless, the most powerful predictor of satisfaction with life was place attachment, with a very relevant impact (more than twice than age). This result stresses the importance of the affective tie with the living environment also in the urban context. In line with the World Health Organization approach "Health in all Policies," present findings demonstrate once more the key role played by the context in fostering well-being and psychological health. Based on the idea that the origins of disease are often outside the health sector, as classically understood (Annett & Nickson, 1991), interest is growing in developing an environmental analysis focused on the impact of place on health and well-being (Atkinson,

Fuller, & Painter, 2012; Macintyre & Ellaway, 2000; Rollero & De Piccoli, 2010a). The importance of territorial communities – or at least some aspects of them – for quality of life of individuals has to be an important subject of study and interest for both researchers and public healthcare providers. Indeed, this perspective could be of great interest for policy-makers in the field of social and health service management, because if it is clear that individuals' life satisfaction differs according to the characteristics and dimensions of the environment, this could affect the organization of health and social services, so that health policies can be tailored to specific needs and life situations (Fassio, Rollero, & De Piccoli, 2013).

The second aim of the study was to clarify the role of different psychosocial variables assessing the quality of the relation between participants and the neighbourhood: i.e., the affective tie with the place (place attachment) and the identification with the social group of residents (identification with inhabitants). Both variables were strongly correlated with satisfaction with life and between them. Yet, the regression analysis allowed showing that just place attachment had a significant impact on satisfaction with life, whereas the identification with the inhabitants had no direct influence. From a theoretical point of view, the cognitive (i.e., identification) and the affective dimensions (i.e., place attachment) reflect the multiple ways that people experience their environment both as a place and as a social group of inhabitants. In other words, place attachment and identification are part of the same overarching self-in-place psychological framework with emotional and cognitive aspects all contributing to the individuals' bonding to an environment (Rollero & De Piccoli, 2010a). These two constructs provided similar outcomes for what concerns the perception of the environment (Rollero & De Piccoli, 2010a). However, the present study demonstrates they lead to distinct results when life satisfaction is considered. We can argue that the affective bond with the environment is more relevant than the cognitive dimension for individuals' well-being. Nevertheless, further research is needed to explore this issue.

The ecological approach considers the interdependence between individuals and contextual systems where people live in. To promote well-being, such contexts have to allow structural conditions to foster healthy lifestyles (Amerio, 2000; Bertini, 2012; Zani & Cicognani, 2000). According to Annett and Nickson (1991), the origins of disease are outside the health sector: for this reason, it is necessary considering the psychosocial processes linking the human beings to their environment and affecting wellbeing and quality of life.

References

Altman, I., & Low, S. M. (1992). *Place attachment. A conceptual inquiry.* New York: Plenum Press.

Amerio P. (2000). *Psicologia di comunità*. Bologna: Il Mulino.

Annett, H., & Nickson, A.J. (1991). Community Involvement in Health: Why Is It Necessary. *Tropical Doctor*, 21, 3-5.

Atkinson, S., Fuller, S. & Painter, J. (Eds.) (*2012*). *Wellbeing and Place*, Farnham: Ashgat.

Bertini, M. (2012). *Psicologia della salute.* Milano: Raffaello Cortina.

Bonaiuto, M., Fornara, F., Aiello, A., & Bonnes, M. (2002). La qualità urbana percepita. In M. Prezza, & M. Santinello (Eds.), *Conoscere la comunità* (pp. 133–160). Bologna: Il Mulino.

Cummins, R. A. (2005). Moving from the quality of life concept to a theory. *Journal of Intellectual disability research*, *49*(10), 699-706.

De Piccoli, N. (2014) (a cura di). *Salute e qualità della vita nella società del benessere*. Roma: Carocci.

Diener, E. (2006). Guidelines for national indicators of subjective well-being and ill-being. *Applied Research in Quality of Life, 1*(2), 151-157.

Diener, E. D., Emmons, R. A., Larsen, R. J., & Griffin, S. (1985). The satisfaction with life scale. *Journal of personality assessment, 49*(1), 71-75.

Dzuka, J. (2012). Subjective evaluation of quality of life: definition, measurement, and starting points for further research. *Ceskoslovenska Psychologie, 56*(2), 148-156.

Engel, G. L. (1977). The need for a new medical model: a challenge for biomedicine. *Science, 196*(4286), 129-136.

Fassio, O., Rollero, C., & De Piccoli, N. (2013). Health, quality of life and population density: A preliminary study on "contextualized" quality of life. *Social Indicators Research, 110*(2), 479-488.

Fried, M. (2000). Continuities and discontinuities of place. *Journal of Environmental Psychology, 20*, 193–205.

Gattino, S., & Tartaglia, S. (2015). The effect of television viewing on ethnic prejudice against immigrants: A study in the Italian context. *International Journal of Intercultural Relations, 44*, 46–52.

Gattino, S., Rollero, C., & De Piccoli, N. (2015). The influence of coping strategies on quality of life from a gender perspective. *Applied Research in Quality of Life*. DOI 10.1007/s11482-014-9348-9.

Hernandez, B., Hidalgo, M. C., Salazar-Laplace, M. E., & Hess, S. (2007). Place attachment and place identity in natives and non-natives. *Journal of Environmental Psychology, 27,* 310–319.

Hidalgo, M. C., & Hernandez, B. (2001). Place attachment: conceptual and empirical questions. *Journal of Environmental Psychology, 21*, 273–281.

Kawachi, I., & Berkman, L. F. (Eds.). (2003). *Neighborhoods and health.* New York: Oxford University Press.

Lent, R. W. (2004). Toward a Unifying Theoretical and Practical Perspective on Well-Being and Psychosocial Adjustment. *Journal of Counseling Psychology, 51*(4), 482.

Lewicka, M. (2010). What makes neighbourhood different from home and city? Effects of place scale on place attachment. *Journal of Environmental Psychology*, 30, 35-51.

Lewicka, M. (2011). Place attachment: How far have we come in the last 40 years? *Journal of Environmental Psychology*, 31, 207-230.

Macintyre, S., Ellaway, A. (2000). Ecological Approaches: Rediscovering the Role of the Physical and Social Environment, in L. Berkman, J. Kawachi (eds.). *Social Epidemiology*, Oxford: Oxford University Press, 332-48.

Mael, F. A., & Tetrick, L. E. (1992). Identifying organizational identification. *Educational and Psychological Measurement, 52*, 813–824.

Mannarini, T., Tartaglia, S., Fedi, A., & Greganti, K. (2006). Image of neighborhood, self-image and sense of community. *Journal of Environmental Psychology, 26*, 202–214.

Meléndez, J. C., Tomás, J. M., Oliver, A., & Navarro, E. (2009). Psychological and physical dimensions explaining life satisfaction among the elderly: A structural model examination. *Archives of Gerontology and Geriatrics, 48*(3), 291-295.

Moser, G., & Ratiu, E. (2002). Appropriation and interpersonal relationships from dwelling to city through the neighborhood. *Environment and behavior, 34*(1), 122-136.

Rollero, C. (2013a). Men and women facing objectification: the effects of media models on well-being, self-esteem and ambivalent sexism. *Revista de Psicologìa Social, 28*, 373-382.

Rollero, C. (2013b). The town in my mind: How place attachment and identification are linked to place perception. *Estudios de Psicología, 34*(3), 309-314.

Rollero, C., & De Piccoli, N. (2010a). Place attachment, identification and environment perception: An empirical study. *Journal of Environmental Psychology, 30*(2), 198-205.

Rollero, C., & De Piccoli, N. (2010b). Does place attachment affect social well-being?. *Revue Européenne de Psychologie Appliquée/European Review of Applied Psychology, 60*(4), 233-238.

Rollero, C., Gattino, S., & De Piccoli, N. (2014). A gender lens on quality of life: the role of sense of community, perceived social support, self-reported health and income. *Social Indicators Research, 116*(3), 887-898.

Rollero, C., Glick, P., & Tartaglia, S. (2014). Psychometric Properties of Short Versions of the Ambivalent Sexism Inventory and Ambivalence Toward Men Inventory. *TPM - Testing, Psychometrics, Methodology in Applied Psychology, 21*, 149-159.

Rollero, C., & Tartaglia, S. (2013). Men and women at work: The effects of objectification on competence, pay, and fit for the job. *Studia Psychologica, 55*, 139-152.

Scannel, L., & Gifford, R, (2010). Defining place attachment: A tripartite organizing framework, *Journal of Environmental Psychology*, *30*, 1-10.

Sirgy, M. J. (2011). Theoretical perspectives guiding QOL indicator projects. *Social Indicators Research, 103*(1), 1-22.

Tartaglia, S. (2009). A comparison between theoretical and naïve definitions of community. In M.F. Hindsworth & T.B. Lang (Eds.), *Community Participation and Empowerment* (263-276). Hauppauge (NY) : Nova Science Publishers.

Tartaglia, S. (2013). Different Predictors of Quality of Life in Urban Environment. *Social Indicators Research, 113* (3), 1045-1053.

Tartaglia, S. (2014). Alcohol consumption among young adults in Italy: The interplay of individual and social factors. *Drugs: education, prevention and policy, 21*(1), 65–71.

Tartaglia, S., & Rossi, M. (2015). The local identity functions in the age of globalization: A study on a local culture. *Community Psychology in Global Perspective, 1*(1), 105-121.

Twigger-Ross, C. L., & Uzzell, D. L. (1996). Place and identity process. *Journal of Enviromental Psychology, 16*, 205–220.

Uzzell, D., & Moser, G. (2006). Environment and quality of life. *Revue Europeenne de Psychologie Appliquee/European Review of Applied Psychology, 56*(1), 1-4.

Wen, M., Hawkley, L.C., & Cacioppo, J. T. (2006). Objective and Perceived Neighborhood Environment, Individual SES and Psychosocial Factors, and Self-Rated Health: An Analysis of Older Adults in Cook Country, Illinois. *Social Science & Medicine, 63*(10), 2575-2590

Zani, B., & Cicognani. E. (2000). *Psicologia della salute*. Bologna: Il Mulino.

In: Urban and Built Environments
Editor: Alexis Cohen
ISBN: 978-1-63483-067-6
© 2015 Nova Science Publishers, Inc.

Chapter 2

WALKING FOR HEALTH: CONTRADICTIONS BETWEEN ENVIRONMENTAL INDICATORS FOR A WALKABLE NEIGHBOURHOOD AND HUMAN BEHAVIOUR

Stefan E. Fina[*]
Institute of Regional Development Planning, Universität Stuttgart,
Stuttgart, Baden-Württemberg, Germany

ABSTRACT

The topic of walkable neighbourhoods is particularly prominent in the literature on Sustainable Cities in the Anglo-Saxon world. There is convincing evidence that segregated, spread-out land uses typical of urban sprawl (e.g., long distances between residential, business, industrial, shopping and recreational functions) have led to an amount of car reliance that impacts drastically on the opportunities, but also the willingness of urban residents to undertake their trips on foot. There is likely to be a health aspect, with numerous studies showing that people who live an active lifestyle and walk a lot are less exposed to a number of diseases and health-related effects that are usually attributed to an inactive, sedentary lifestyle (e.g., non-communicable diseases like obesity, coronary heart diseases and type-2 diabetes). Spatial indicators like the Walkability Index or the Walkscore have been developed in this

[*] Email: stefan.fina@ireus.uni-stuttgart.de

context and help geographical and business analysts to highlight problem areas, or vice versa to stimulate demand advantages for walkable neighbourhoods on the real estate market.

Despite these efforts, the number and distance of walking trips in the Anglo-Saxon world remain amongst the highest in the world. Comparisons with mainland Europe show that urban residents walk and bike a lot more throughout all social classes, especially in parts of Europe where the climate is not always favourable of active lifestyles (The Netherlands, Scandinavia, Germany). In this context, recent adaptations of the Walkability Index and Walkscore for the City of Stuttgart, Germany, have shown that the urban structure approach they are based upon does not necessarily explain the amount that people walk very well. In a comparison with a large household survey database, people who walked less did not necessarily live in the least walkable areas and vice versa. The conclusion was that (1) there must be either other environmental forces at work that are not captured by the aforementioned indicators, or that (2) cultural and behavioural differences are predominant and environment does not play that much of a role after all.

This article systematically looks at a more comprehensive set of environmental conditions and personal characteristics in the neighbourhoods of survey respondents, with the aim to identify variables that play a specific role in the European context. It briefly reviews the literature on behavioural determinants for walking and gives an overview over current concepts to measure walkable neighbourhoods and highlights their shortcomings. In the methodological chapter, multi-level regression analysis is used to control the survey dataset for personal precursors for walking like age, employment, health status, and transport mode options. Based on these results, the article determines conceptual gaps in current walkability assessments and analyses what other parameters of urban form than the ones used by current walkability assessment tools have explanatory power, and what factors can be attributed to behavioural differences. The final chapter discusses what city planners and architects can learn from the results in order to deliver walkable neighbourhoods that stimulate walking for health more effectively.

BACKGROUND

It is a well-researched fact that the population in many parts of the Western world live unhealthy lifestyles when it comes to physical exercise in their daily activities. The need to walk to and from destinations has decreased with infrastructure development and its accompanying transport options in modern times, especially in sprawling cities with a high perceived and actual dependency on car travel. It is for many reasons that trips on foot are decreasing, and a decreasing fitness level of the population can either directly or indirectly be attributed to this trend. The amount of walking that is estimated to maintain the personal fitness becomes manifest in WHO recommendations and many local health programmes (10,000 steps to work), the popularity of health apps that help to monitor individual physical activity, and some radical suggestions to rebuild cities to force people to walk more (see for example Bucksch und Schneider, 2014).

At the same time the individual is exposed to a number of influencing factors that may work against a balanced amount of walking as a daily activity. Researchers observe increasing number of personal trips that cover distances between origins and distances that not everyone can cover easily by foot. Examples for such tendencies can be observed in the increasing flexibility that many jobs require from employees, as can be seen by a massive increase of car, train and plane trips for business reasons. In such cases, there is limited dispute about the necessity of such trips and the argumentation why not even shorter trip legs can be made on foot (time constraints, appearance in meetings, etc.) is widely accepted. One could conclude from these observations that modern lifestyles require a level of flexibility that hinders many people to plan their trips with the limited flexibility (walking) or uncertainties and risks (biking) of active mobility modes (see for example Jack und McCormack, 2014; Glazier et al., 2014).

In other mobility areas like trips to school, where an increasing number of parents drive their children to school for reasons like safety, different worldviews of good parenting collide with the principles of sustainable mobility (Lavoie et al., 2014). Although controversial, the essential right to choose the most convenient transport option remains undisputed in democratic societies. And in some societies there are either cultural, religious or climatic restrictions to walking that are not necessarily part of the research agenda of transport and city planners. However, the rising awareness that walkable neighbourhoods and liveable communities are closely related has inspired many policy-makers to tackle travel behaviour through travel demand management programmes that are being implemented on various levels, from road pricing initiatives to voluntary community development initiatives (e.g., walking school busses). Community walkability audits are being undertaken and best practice design standards for a pedestrian-friendly environment are on the bookshelves of many transport planners and engineers around the globe. The success of such initiatives is mixed at best. Apparent success measured by an increase in the number of trips made by foot is often offset by an actual stagnant or decreasing mode share due to the general increase of trips (Gärling und Schuitema, 2007).

Comparative research by Bühler und Jung, 2013, shows that city environments exist especially in Northern Europe (for example Amsterdam and Copenhagen), where the share of active trips is by order of magnitudes higher than for cities in North America (particularly in Washington DC where they conduct most of their research). Their conclusion is that not only the infrastructure and city layout plays a decisive role in the amount people walk, but also a certain planning culture where authorities do not hesitate to implement unpopular measures like the removal of parking in the inner city or road pricing rather than voluntary measures. The public debate in these cities seem to be less prejudiced than in other parts of the world where policies to educate people and change their mobility behaviour are all too easily labelled as "social engineering" or "carrot and stick" with all its negative connotations (Meyer, 1999).

In a more conceptual approach to assess walkability, Ewing und Cervero, 2010 have established the four 'D' model around the city structure criteria (a) diversity (= mix of land uses in a neighbourhood, many destinations to walk to), (b) density (= level of density in the urban/community space with a higher likelihood of presence and close proximity to different destinations, (c) design (= aspects like the availability and layout of foot- and bikepaths, the connectivity of roads and areas with highly attractive community spaces like squares and popular parks, and (d) destination (= distance and accessibility of important destinations for daily errands, recreation and entertainment in the local neighbourhood area. And lastly (e) distance to travel (= the distance from residential origins to workplace destinations or walkable transfer points like public transport stops. These considerations have been quite influential in measurement approaches that have subsequently been developed. The next chapter gives an overview over these methods.

Methods to Measure Walkability

There are a number of methods available to assess the walkability or pedestrian-friendliness of urban areas or neighbourhoods. Bucksch und Schneider, 2014, differentiate these methods into (a) audit-based tools that require experts to assess certain level of service conditions of the streetscape like safety obstacles, cleanliness, street lighting, width and quality of footpaths, etc. and possibly combine these with surveys where pedestrians give their opinions on the attractiveness of the environment to walk. And (b) into methods that derive their information from secondary data sources and adopt them for indicator calculations (i.e., cadastral data, facilities databases, remote sensing land use information, etc.). The advantage of the latter is clear: audits require resources to collect information with efforts that often allow implementation only for certain selected neighbourhoods on a project basis or with a limitation on the number of collected attributes and resident survey coverage or sample size (Scanlin et al., 2014; Giles-Corti et al., 2015). Methods based on secondary information can draw from existing databases and cover larger areas with less effort. The disadvantage is that they are not likely to generate the same level of information than specifically designed audits. There is a reliance on datasets that often have to be combined and harmonized from several databases, with some inherent risks of reducing the depth of information throughout the process.

At the same time, the outcome in the form of indicators of walkability allow for standardized comparisons of the built environment over and between larger regions, which can prove to be a valuable asset to the methodological planning instruments of researchers and authorities alike. Over the past years, much research has been dedicated to the development of such indicators at the Institute of Regional Planning at the University of Stuttgart, three of which will be further explained in the following - with an aim to assess the attractiveness of residential and mixed use neighbourhoods for active modes of transportation in general, and walking in particular:

- The Walkabiliy Index is a tool developed by the IPENZ organisation (Dobesova und Krivka, 2012) based on the results of their research efforts. Its aim is to work out the general attractiveness of a study area for walking based on generalized spatial indicators that have been found to be influencing people's willingness to walk (road connectivity, residential density, mixed use structures, retail floor space).
- The WalkScore[1] focuses on the accessibility of destinations for frequent activities within walking distances. It is a private business development that is used to assess the liveability of a neighbourhood, and a measurement relevant to the real estate industry especially in the United States. As such, the facilities used prioritize shopping and

[1] See www.walkscore.com, last accessed 29 April 2015

entertainment facilities in the assessment over social and recreational facilities. Adaptations like the bikescore have a similar aim, with enhancements toward the suitability of infrastructure and street equipment for safe biking to amenities relevant for cyclists.

- The car dependence study undertaken at the University of Stuttgart is a novel concept that aims to assess neighbourhood characteristics that force residents to use the car. One part of that assessment ("non-motorized car dependency") looks at the pedestrian accessibility of facilities with a focus on shopping as well as medical and social institutions (general practitioners, pharmacies, kindergarten and elementary schools) which is used here as a proxy for walkability.

Some smaller area implementations for the City of Stuttgart (Walkability Index, Walkscore) and a German publication on the car dependency in the Region of Stuttgart have already been published (Reyer et al., 2014; Siedentop et al., 2013). The main focus then was to explain methodological and implementation issues and deliver some preliminary interpretations, including some limited validation efforts against a travel behaviour database for the walkscore and the walkability index. This article goes one step further and extends the implementation of these three measures to the Region of Stuttgart, standardizes the measurements in terms of geographic units, and systematically tests the results against behavioural aspects of over 9000 respondents of a travel behaviour survey.

MEASURES OF WALKABILITY IN THE STUTTGART REGION

The Stuttgart Region is located in the Southwest of Germany (see figure 1). With a population of about 2.7 million it is one of the larger urban agglomerations in Germany with a thriving economy mainly based on the automobile industry and a growing service sector. It is the state capital of Baden-Württemberg with a stable inner city population of just under 600.000 inhabitants and a sprawling region with growing secondary centres like Ludwigsburg, Esslingen, Böblingen, Sindelfingen and Waiblingen. The urban-rural fringe is characterized by historic small towns and villages with a variety of modern farming practices towards the South, West, and North, and a distinct cultural landscape along the slopes of the Swabian Alb in the Southeast. From a regional planning point of view the region is well known for its polycentric setup with strong transport links between the inner city and the sub-centres, serviced by a highly-frequented multi-modal network of motorways and trains (high-speed Inter-City-Express, regional trains, interurban trains, trams) (see for example Siedentop et al., 2014; Fina, 2013; Region Stuttgart, 2015).

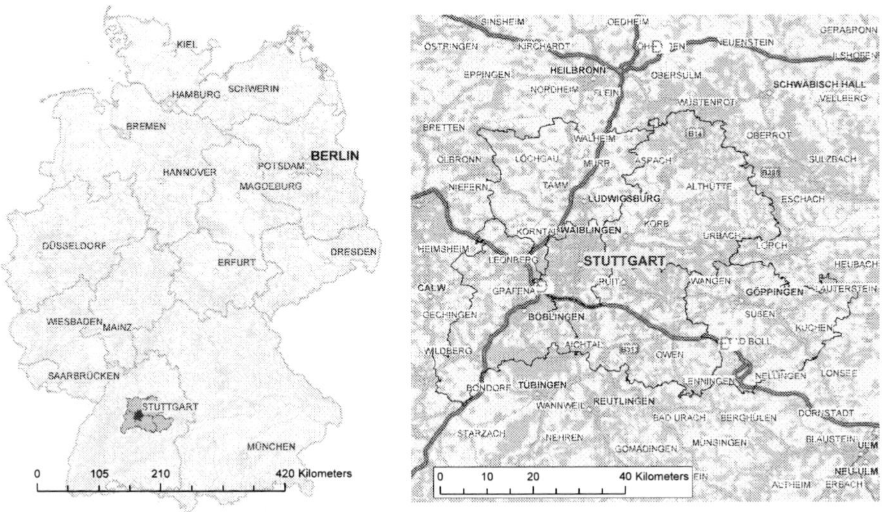

Figure 1. The location of the Stuttgart Region within Germany (left) and close-up of the Region (right). (Data Source: GeoStreet+, infas Geodaten 2010).

Table 1. Modal split in the Stuttgart Region (trips made by persons older than 6 years)

Transport mode	Trips per year in millions		Trips in %	
	All days	Mo-Fr	All days	Mo-Fr
Walking	655.1	483.8	23.6	22.9
Cycling	188.2	152.9	6.8	7.3
Car driver	1205.5	945.9	43.8	44.8
Car passenger	381.3	229.0	13.7	10.9
Public Transport	346.4	298.2	12.5	14.1
Other	1.6	1.1	0.1	0.1
Total	2778.1	2110.9	100.0	100.0

Source: translated from Verband Region Stuttgart, 2011, p. 77.

A study conducted by the regional administrative body, responsible for interregional public transportation, has collected household survey information from about 13 thousand survey respondents and used this as the base for the total modal split figures shown in table 1 (Verband Region Stuttgart, 2011).

In comparison to data collected from various sources published on Wikipedia[2] (Urban Audit for Europe, US Census for United States, ABS Census for Australia), the City of Stuttgart has a rather low level of people walking compared to other (but mostly larger) European city regions, but a much higher percentage than any city region in the United States or in Australia. In general the data presented on Wikipedia suggests that (a) there is a region difference between European cities and cities from the United States or Australia, and (b) there is a size difference in the sense that people in larger (European) cities tend to walk more and use public transport more frequently.

Please note that the validity of this comparison depends somewhat on the definition of city regions in terms of their boundaries ("how much hinterland is part of the city?"), and methodology that was used to generate information (estimates vs. measurements). In that sense, the comparison of European cities on the European Platform on Mobility Management (EPOMM[3]) is more robust since they are using similar spatial classifications with standardized inputs and outputs. Here, walking in Stuttgart has a rather low percentage, public transport use is average, but car use is fairly high.

Drilling deeper, figure 2 shows the spatial variation of the walking mode throughout the region, covering 624 transport zones. It is based on the transport model the region has developed and calibrated against the data from the survey mentioned above. The percentages shown actually aggregate walking and cycling to a combined "slow mode." Unfortunately this cannot be output as walking and cycling separately, so that any further interpretations of this data account for both modes.

The map clearly shows that people walk and cycle more in the centre and of course also in the sub-centres. With some knowledge about the local geography, this picture is not surprising, but it delivers a sound validation base for the performance of walkability indicators for the same region. On this background, the next question to answer is whether walkability indicators can be used as predictors for higher mode shares of walking (and cycling).

---

[2] http://de.wikipedia.org/wiki/Modal_Split#St.C3.A4dte_in_Deutschland, last accessed 29 April 2015.

[3] http://www.epomm.eu/tems/cities.phtml, last accessed 29 April 2015.

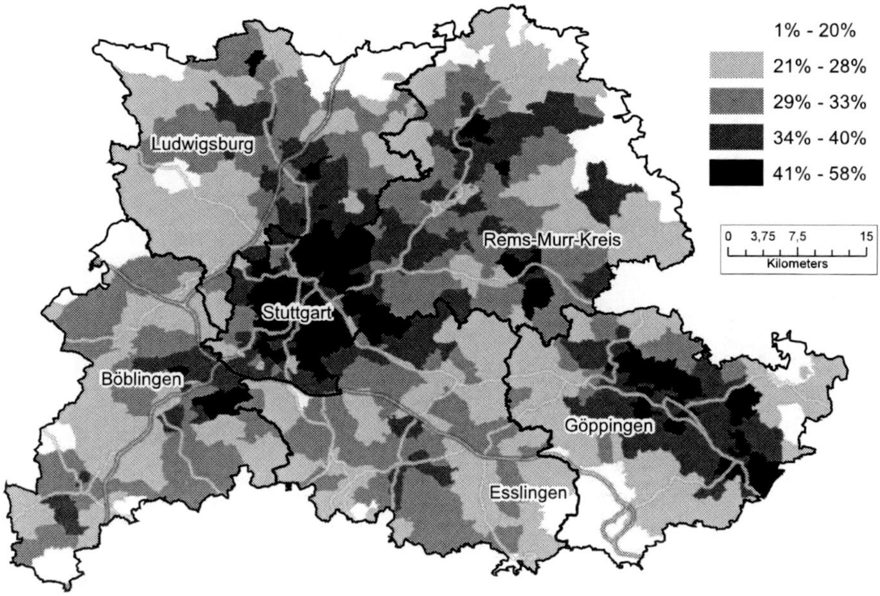

Figure 2. Share of walking trip in the Stuttgart Region (Data: Visum transport model, Verband Region Stuttgart, 2012).

For this purpose, the collection of maps and snapshots of neighbourhoods presented in figure 3 gives an initial overview over the three indicators employed for this study. The design of this figure is meant to

- Indicator variations: illustrate the indicator with close-ups of neighbourhoods that portray a minimum and maximum of an indicator's value range, and also showing the main influencing factors;
- Generalized indicator representations for the whole region, using a heat mapping procedure to extract the regional patterns from the neighbourhood values of each indicator[4].

[4] Kernel Density tool with a Gaussian curve and 2000 meter bandwidth (= maximum walking distance in the Walkscore implementation); software used: ArcGIS 10.2.2

Walkability Index	Heat map
Values rely on the levels of population density, street connectivity, land use mix and presence of retail floorspace (facilities in the maps below). Maximum values tend to dominate in well-serviced dense inner city neighbourhoods, minimum values in rural farming communities, in this case with a road system in hilly topography.	The spatial pattern follows a population density and urban-rural divide, clearly highlighting the inner cities with highest walkabilities. Values decrease along the urban-rural continuum towards the periphery.

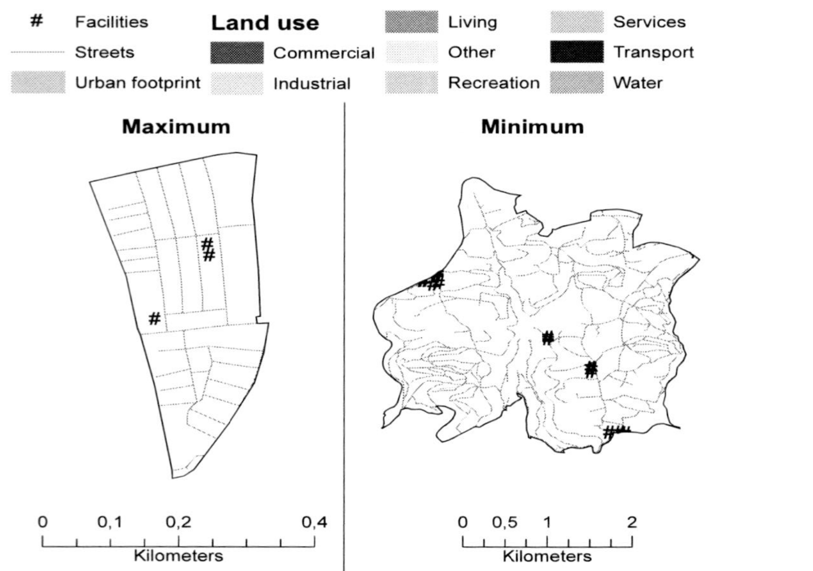

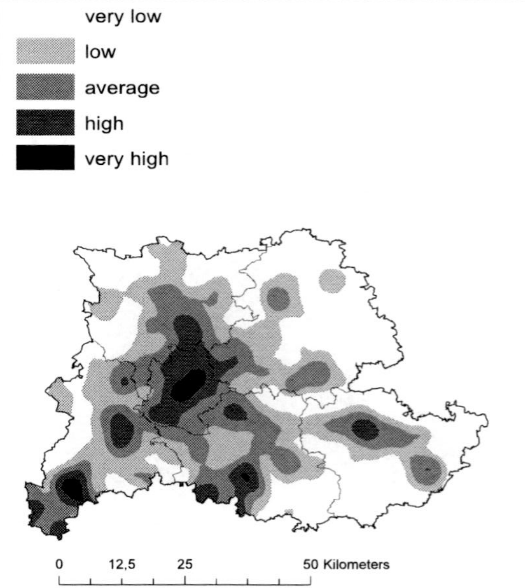

Figure 3. Snapshot of the Walkability Index and its maximum and minimum values amongst Stuttgart neighbourhoods (left), and a generalized overview (right) (Data: GeoStreet+, cadastral database).

Walkscore	Heat map
Values rely on the walking distances towards selected facilities, with decreasing weightings for a choice of grocery stores, shopping, restaurants, cafes (several), and bookstores, banks, entertainment, schools and parks (just one). The closer the facility is to the place of residence along the walking network, the higher the score. Minimum values are therefore in remote areas far away from any facility, maximum values are where almost all errands can be reached easily in very close proximity to the point of residence.	The variation in the Stuttgart Region corresponds with the density of facilities, which is naturally larger in the most populated areas with the largest customer base. Highest walkabilities are therefore also in inner cities and decrease along the urban-rural continuum towards the periphery.
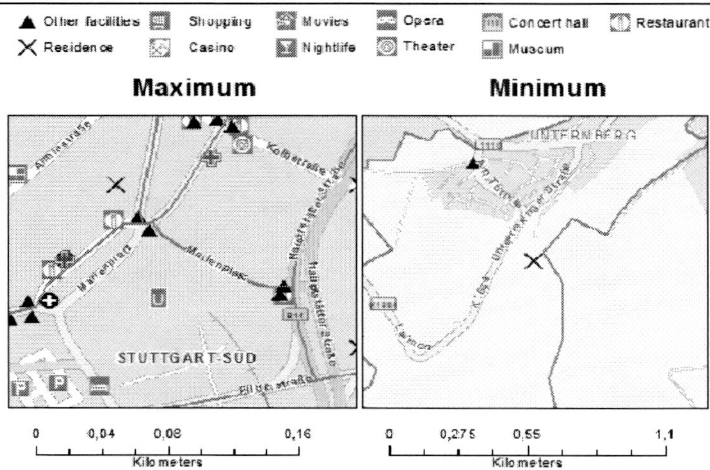	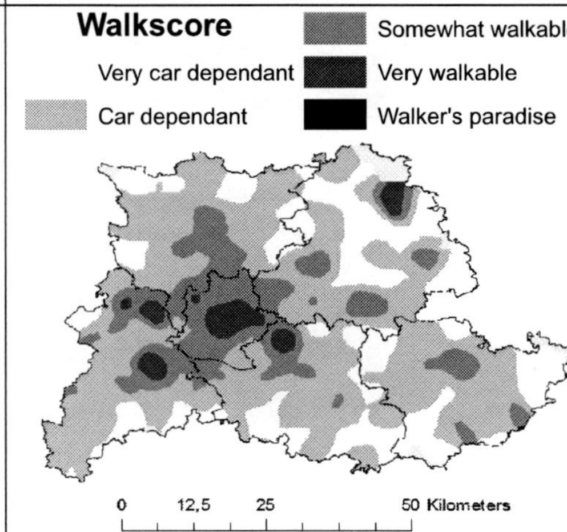

Figure 4. Snapshots of the Walkscore and its maximum and minimum values amongst Stuttgart neighbourhoods (left), and a generalized overview (right) (Data: GeoStreet+, cadastral database).

Car dependency – non-motorized transport	Heat map
Values here are strongly influenced by (a) neighbourhood aspects like the accessibility of medical and social facilities by non-motorized transport, and (b) the competitiveness of public transport alternatives to car travel. Minimum values predominate in inner-city locations, maximum values can be found in remote rural locations.	The pattern here is more centralized, highlighting the Stuttgart centre and its immediate surrounds and few sub-centres. Values decrease with increasing distance to the centres, which could be due to medical and social facilities being more concentrated.
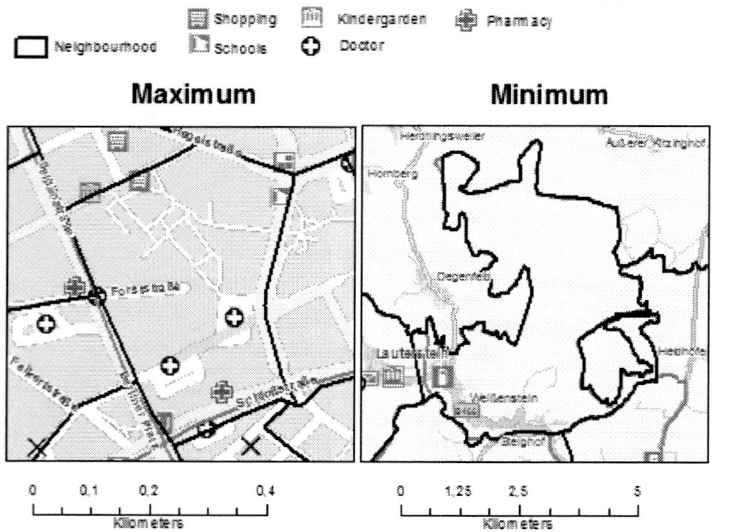	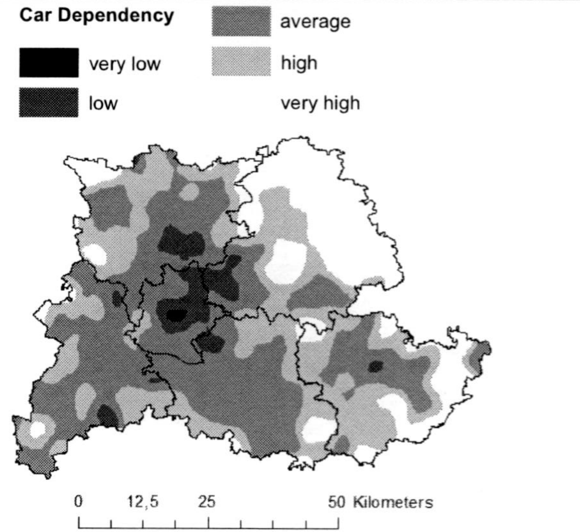

Figure 5. Indicators of Walkability for Region Stuttgart (Data sources: cadastral database ALK 2010, GeoStreet+ 2010, YellowPages facilities 2011, Visum Transport Model University of Stuttgart, , 2012).

METHODOLOGICAL ISSUES

There are a number of methodological drawbacks either due to data or accuracy limitations that need be mentioned. The author's overall assessment is that these drawbacks do not affect the meaningfulness of the indicator message substantially, at least not at an order of magnitude that would play a role compared to their conceptual limitations – which will be further elaborated upon in the following.

Methodological issues around the Walkability Index are mainly around the level of detail that the data can deliver. In the implementation presented here, the highly detailed street network (GeoStreet+, based on Teleatlas/ TomTom navigation data), the cadastral database used for computing the land-use mix and retail floorspace deliver highly detailed information on an area-wide basis. From a conceptual point of view, reviewers in previous articles have criticized the inclusion of commercial area in the land-use mix since they could potentially not include any relevant destinations for residents, and the lack of qualitative information on the street network (quality and safety of footpaths).

In contrast to the other indicators, walkscore calculations were not applied to neighbourhood zones, but to geo-located addresses of survey respondents. The process of geocoding used the geographic coordinates in the mobility database with an accuracy drawback: Residential coordinates outside the City of Stuttgart are only given with two decimal digits. At the Stuttgart latitudes and longitudes (around 9.2/48.2°at the centre), deviations in a Gauss-Krüger Zone 3 projection could be up to 200 meters for the latitude and up to 900 meters for the longitude. This information was taken from sample measurements using (a) known higher accuracy residential points from within the city, (b) reducing their accuracy to two digits, (c) measuring the projected length between the (a) and (b). The network distances were calculated along the GeoStreet+ network, with no further information on footpath quality. From a conceptual point of view, the selected facilities sourced from YellowPages information can be assumed to be close to complete, but they do not contain further information on the services offered and therefore on their actual attractiveness for customers.

With regards to the car dependency study, the same street network with its lack of footpath information was used. The medical facilities used from YellowPages data can be assumed to be fairly complete and also allow for a distinction of professions (i.e., general practictioner). Shopping facilities were not differentiated by attractiveness (i.e., a small bakery has potentially the same attractiveness than a larger supermarket), and the weighting of the facilities for a subsequent classification in terms of car dependency is based on theoretical assumptions using statistical quantile thresholds, rather than empirically-derived curves (see Siedentop et al., 2013 for a detailed explanation).

With these limitations in mind, the next step in the analysis chain is to compare the indicator results against the walking mode share derived from survey respondents.

VALIDATION OF WALKABILITY ASSESSMENTS

The validation of the walkability measures follows a research design that distinguishes between the neighbourhood and household level. Initial attempts to utilize statistical multi-level analysis have not yielded any significant results. As an alternative, a stepwise procedure using (a) linear regression analysis and geographically weighted regression on the neighbourhood level have yielded some clues as to what variables are useful and which ones might be missing, and (b) linear regression on the household level including additional dependent variables complement the research design.

Validation of Walkability Indicators against Modal Split Values on the Neighbourhood Level

The validation of the results presented above starts with a linear regression model[5] in figure 6, taking the share of slow mode trips from the transport model as the variable to explain, and the three indicators explained above as the variables that we suppose are capable of predicting the share of slow modes. Since this share is a modelled value that is calibrated against the survey database, it is here defined as the "reality of walking" that can be tested statistically for each of the three indicators explained above - or their combination. Issues of collinearity might be present between the walkscore and the non-motorized car dependency, since they both work with accessibility to shopping facilities as a component of their final value.

[5] Using the SPSS linear model fit method with no bootstrap, the share of slow mode trips as the dependant variable, and the walkability index, walkscore and non-motorized car dependency as the independent variables.

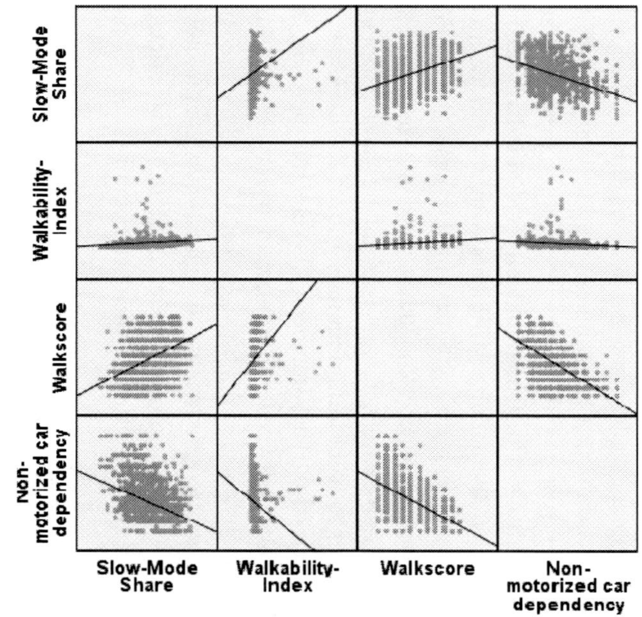

Coefficientsa					
	Unstand. Coeff.		Stand Coeff		
Model	B	Std. Error	Beta	t	Sig.
(Constant)	.347	.006		54.4	.0
Walkability Index	.002	.001	.074	3.6	.0
Walkscore	.001	.000	.277	11.1	.0
Non-motorized car dependency	-.092	.010	-.212	-8.7	.0
a. Dependent Variable: Slow mode share					

Figure 6. Statistical validation using linear regression analysis (right, adjusted r^2=0.213) and matrix scatterplots with a fit line.

However, they differ in the in- or exclusion of other facilities so that including both to begin with is acceptable. It also has the advantage that it can be spatially harmonized with the indicator database, taking each of the observation units from the transport model and spatially overlaying it with the units the other indicators have been calculated for (n=2128)[6].

With a rather low explanation value (r^2=0.213) the model shows almost no explanatory power for the walkability index, but some explanation value for the walkscore (positively) and the non-motorized car dependency (negative, i.e., areas with higher shares of slow mode trips have less car dependency, see the highlighted standard coefficients on the right hand side of figure 6). The linear fitting lines on the left side in the figure support the interpretation that the walkability index does not work very well as a predictor of a higher share of slow mode trips, but that the walkscore and the non-motorized car dependency indicators do to some degree.

In the next attempt to extract more explanatory power from this comparison, geographically weighted regression was used. This method conducts regression analysis using an iterative moving window over the observation points, thus only comparing values within a fixed geographic extent at a time. The reason why this method is quite promising in this context is because in a regional context we have a large variation of settlement structures along the urban-rural continuum that naturally vary in terms of infrastructure, housing types, land use) that are just too diverse to be combined in one statistical analysis. The Geographically weighted regression cuts out some of this “noise” by calculating coefficients within a bandwidth of a certain distance or a fixed number of neighbours. In this case, 30 neighbours were used, mainly because the observation units vary in size (more and smaller units in centres and sub-centres, larger ones in the rural periphery) and the neighbour method grabs a more harmonious subset for each iteration than a fixed distance could do. The number 30 selected here is an average neighbourhood[7] number between the number of observation points necessary to capture homogeneous areas within the total set of 2128 data records.

As expected the geographically weighted regression delivers more explanatory power with an adjusted r^2-value of 0.71. This means that the spatial indicators predict ok within a local environment of 30 neighbouring

---

[6] Not all observation units had valid values due to missing data. In such cases, proximity analysis was used to spatially harmonize the database on the basis of the nearest observation point (e.g., area unit centroids).

[7] Neighbourhoods have between 500-1000 households each, based on former election wards.

observation units. If the walkability index is excluded, the r^2-value goes up to 0.78.

Figure 7 shows the resulting residual map with the z-values (i.e., the standard deviations of each observation point from the regression line). This illustration helps to see where local predictions deviate most from the fitted model line, i.e., where it is likely that either the observed data is for unknown reasons skewed or certain variables are missing.

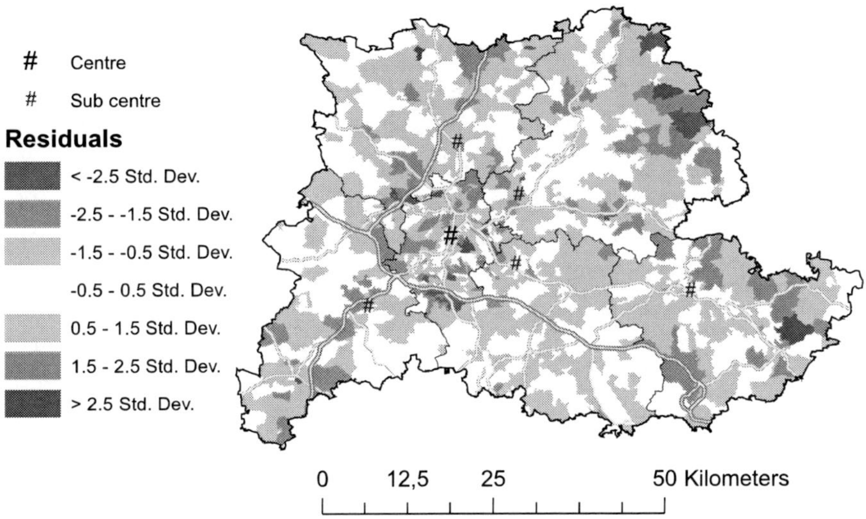

Figure 7. Statistical validation using linear regression analysis (right, adjusted r^2=0.213) and matrix scatterplots with a fit line.

With some local knowledge it appears that negative deviations (i.e., in these places people walk less than the model predicts) lie in the suburban rings around centres and sub-centres, and positive deviations (i.e., people walk and cycle more than predicted) are within or in close proximity to city centres.

Validation of Walkability Indicators against Surveyed Person Data on the Household Level

With this interpretation in mind the search for possibly missing dependent variables resulted in two more candidates being included in the analysis. The first one is public transport accessibility, using the assumption that public transport users frequently walk more or less distances to their next stop (see for example Zwald et al., 2014; Djurhuus et al., 2014 to substantiate this assumption). The other one is slope, which is obviously an additional obstacle for rather unfit people to walk (Frackelton et al., 2013).

The other observation is that both of these additional variables can be expected to be more relevant on a household level than a neighbourhood level. Distance to the nearest stop as well as slope are very local conditions that cannot be expressed as average values for a neighbourhood zone. For this purpose, the survey dataset was complemented with public transport accessibility data, which was weighted by the number and quality of services that operate at the nearest stop. Weights were applied in increasing order to bus, tram, local and regional trains, and intercity train stations, and multiplied with the number of services for each of these categories. The threshold for accessibility was set to 2.5 km, beyond which people are not assumed to walk regularly. And finally, the values of three stations were summarized into a final value, so that the attractiveness of a highly serviced station is not ignored if a local bus stop happens to be nearer to the place of residence. With this method, the attractiveness of each stop was given effect in the philosophy of a gravity model without having to set up a complete calibrated model itself (which would have been outside the scope of this study). Similarly, data on elevation was extracted from a 30 meter digital elevation model provided by the state environmental institute (LUBW - Landesanstalt für Umwelt, Messungen und Naturschutz Baden-Württemberg).

Figure 8 shows that public transport attractiveness does in fact show a positive correlation with higher shares of walking (lower centre on the left hand side), but that most of this effect is caused by a skew in the data with few but valid outliers (the most important stations in the region). As for slope, the effect is very low but the tendency supports the hypothesis that people living in hillier environments walk less (lower left). On the right hand side, the figure shows the effect of these two variables in a linear regression model with other personal or household variables that have been identified as relevant after tests for collinearity. They include dummy variables for expressing if a certain property is present or not (e.g., if a person is working or not, if a car is available or not, etc.).

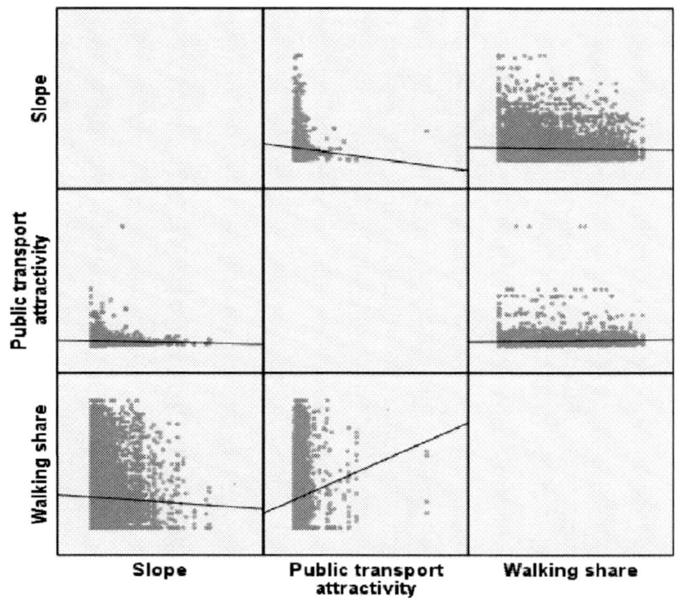

Model	Std. Coeff.
Age	,134
Car Availability	-,102
Public transport ticket	,015
Working	,001
Children between 0 and 6 years	,045
Household income	-,022
Walkscore	,036
Slope	,017
Public transport attractiveness	,003

a. Dependent Variable: Share of walking.

R^2: 0.29, n=9880

Figure 8. Testing two additional variables as explanatory variables for higher shares of walking: public transport accessibility and slope (left) and linear regression modelling results (right).

Overall, the results shown here explain just under 30% of the variation in the walk share (r^2=0.29) and cannot be said to explain the share of walking satisfactorily. With very low explanatory power in general, the model suggests that public transport attractiveness and the slope have almost no effect. The share of walking trips seems to increase with age, and car availability plays a negative role, so that people with cars would have a lesser share of walking trips. There is no effect as to people's working status. Persons in households with small children seem to do more trips on foot, while people with a higher income walk less. Also, the effect of a walkable neighbourhood expressed by the walkscore is small but positive.

Despite these interpretations, the first and foremost result is that there was no good model fit, suggesting that (a) there are either missing variables, (b) that the hypothesis must be rejected, or (c) that the methodological approach is unsuitable to detect inherent patterns. To rule out the latter, multilevel regression modelling was used to control the surveyed walking information for personal level-dependant precursors of walking like age, holding a driver's license, car availability, employment status, or holding of a public transport pass. Other characteristics were fixed on a household level (family with smaller children, the walkscore as the selected neighbourhood walkability measure, the aforementioned variables of slope and public transport accessibility). For such cases, multilevel or hierarchical regression modelling is recommended, using fixed variables for consistent information in a set of information (e.g., household characteristics for each person residing in the household) and random variables for the personal characteristics that vary for each observation. The result of this analysis however, did not produce results worthy of publication, the model formulations did not lead to convergence. This means that no reliable variable constellation could be found that would explain the variance between observed walk shares on the personal level of survey respondents.

LESSONS LEARNED FROM THE MOBILITY DATABASE

The conclusion from the validation efforts described in the previous chapter is that neither neighbourhood nor personal characteristics of survey respondents can explain the variance in the share of walking trips within the Stuttgart region, at least not in a statistically significant way. Simple and multilevel regression modelling attempts have failed to produce reliable explanatory models, although tendencies for variable values were somewhat interpretable. Within the limitations of this failure, a number of concluding remarks are worthy of consideration:

1) The choice of transport mode is a highly individualistic process that does not follow deterministic rules in European city regions like the Stuttgart region.
2) Attempts to define environmental neighbourhood characteristics that would change people's walking behaviour cannot be shown to be of measurable relevance in the Stuttgart region.
3) Compared to the actual mode share for walking, the tested walkability indicators underestimate the amount people walk within cities, and overestimate the same in suburban areas.
4) Additional criteria like public transport attractiveness or slope are conceptually logical, but do not improve the significance of explanation models.

In the following section, these remarks will be further investigated in terms of their meaning for city and public health planners:

The first statement can be explained as a result of the variety of lifestyles that can be found within the regions and the strong level of interconnectedness of the transport system. A look into the trip diaries that were used to establish the mobility database is quite helpful in this respect: Table 2 shows ten random records each for respondents with the highest shares of walking trips, a medium share of walking trips and the lowest shares. Even though just a small sample it appears that there are all expressions the variable value domains present, with a clear pattern that is useful to formulate a hypothesis: the respondents with the lowest shares of walking trips seem to be predominantly workers with cars or public transport passes, with no children in the household, living in rather car dependant neighbourhoods. For people who walk most it's less clear: on average they live in more walkable environments and have a lower car availability, but some live in rather car dependant environments. And between these extremes all variations are possible, patterns are hard to extract and predict. The interpretation is (a) that an increasing number of people have mobility habits that are detached from the immediate surrounds of the neighbourhood, (b) the necessities of trips chains organized within the social network possibly supersedes predictable travel behaviour patterns, and (c) transport system related requirements are not represented in the variable set (e.g., parking availability at the place of destination, mobility impairment, etc.).

With regards to the second statement, the expectation that neighbourhood characteristics are measured accurately by the walkability index, the walkscore or the non-motorized car dependency is already based on several assumptions and theoretical considerations (e.g., choosing sub-indicators like a dense street network, population density, retail floorspace or measuring accessibility of selected facilities as a basis for indicator calculation). It is the nature of such spatial indicators that they only operate as proxies for assumed interrelationships between environmental predicaments and actual human behaviour. It is not the first time that indicator validation fails in this respect (see for example Reyer et al., 2014), but this is not to say that the indicators are meaningless for planners. Even with more accurate and quantifiable audit-based tools one could create a highly walkable environment that people may not use if their personal situation does not allow it. Which, in turn, does not mean that such planning principles should not accompany city development strategies, quite the contrary: walkable communities are highly desirable, but in strongly interconnected city regions mobility patterns are too diverse to evaluate people's behaviours against their neighbourhood environment alone.

With regards to the indicator performance the results show that people in the inner-city actually walk even more than the indicator values show, and that they overestimate the share of walking in suburban environments. The reason for that can only be formulated in a new hypothesis: in the author's point of view, all of the three indicators are conceptually quite weak in the representation of inner city traffic congestion effects and their environmental impacts. The experience of lost time in traffic and being exposed to transport pollution (noise, air) over longer periods of time raises an awareness about the benefits of alternative transport modes that is completely absent in the formulation of walkability indicators (see also Karusisi et al., 2014). Local initiatives like critical mass biking events and political initiatives that support and now have already implemented shared road space initiatives (Stuttgart's Tübinger Straße) are proof of this awareness particularly in the inner cities. The suburbs, in contrast, are in parts the new migration centres where a variety of people without the same level of community cohesion as in the inner city reside, with a higher willingness to accept automobile dependency due to less perceived transport-related environmental stress.

Table 2. Sample records from the mobility database, representing ten random records with high shares of walking trips (top), ten records with medium share of walking trips (center), and ten trips with low shares of walking trips)

Share of walking trips	Age	Wor-king	Car avail-ability	Public transport pass	Driver's license	Young children (0-6 years)	HH income (up to…)	Walk-score	Car depen-dency	Walk-ability index	Walking trips		Other trips		Slope	Public transport attractive-ness
[%]	[years]	[1: yes/2: no]	[1: yes/2: no]	[1: yes/2: no]	[1: yes/2: no]	[1: yes/2: no]	[Euro]	[Points]	[Points]	[Points]	[km]	[n]	[km]	[n]	[°]	[Index]
100%	10	0	0	1	0	0	3500	50	0,10	1,52	11,4	18	0,0	0	9	53,33
100%	36	1	1	1	1	0	1500	42	0,31	1,88	9,4	16	0,0	0	3	20,00
100%	38	1	1	1	1	1	2500	83	0,10	1,80	19,7	33	0,0	0	2	33,33
100%	16	0	0	1	0	0	3500	83	0,00	1,91	24,7	14	0,0	0	14	64,67
100%	79	0	0	1	1	0	2500	67	0,20	-0,18	17,1	14	0,0	0	4	14,00
100%	68	0	0	1	1	0	2500	33	0,27	-0,41	139,8	28	0,0	0	3	33,33
100%	67	0	0	1	0	0	4500	42	0,23	-0,79	25,4	21	0,0	0	9	12,00
100%	50	1	1	1	1	0	1500	75	0,10	1,65	19,0	22	0,0	0	5	46,67
100%	83	0	0	1	0	0	1500	58	0,33	0,48	24,7	18	0,0	0	2	46,67
100%	50	0	0	1	0	0	1500	17	0,33	-1,05	98,3	32	0,0	0	2	48,00
50%	74	0	0	1	1	0	2500	58	0,27	0,87	38,4	22	74,7	22	4	20,00
50%	54	1	1	1	1	0	3500	50	0,33	-0,33	5,7	12	224,2	12	10	80,00
50%	74	0	0	1	1	0	2500	67	0,20	1,12	57,6	14	88,1	14	3	30,00
50%	30	1	1	1	1	1	3000	17	0,33	-1,14	18,0	18	200,0	18	4	30,67
50%	19	0	0	1	0	0	2000	33	0,27	0,06	9,8	11	62,1	11	3	120,00
50%	17	0	0	1	1	0	1500	17	0,33	-0,85	16,7	15	149,0	15	19	24,00
50%	83	0	0	1	1	0	2500	33	0,27	0,06	64,3	7	128,1	7	3	120,00

Table 2. (Continued)

Share of walking trips	Age	Wor-king	Car avail-ability	Public transport pass	Driver's license	Young children (0-6 years)	HH income (up to…)	Walk-score	Car depen-dency	Walk-ability index	Walking trips		Other trips		Slope	Public transport attractive-ness
[%]	[years]	[1: yes/2: no]	[1: yes/2: no]	[1: yes/2: no]	[1: yes/2: no]	[1: yes/2: no]	[Euro]	[Points]	[Points]	[Points]	[km]	[n]	[km]	[n]	[°]	[Index]
50%	18	0	0	1	1	0	4500	8	0,60	2,75	10,3	10	117,6	10	4	50,67
50%	56	1	1	1	1	0	4500	50	0,27	1,44	32,1	15	155,9	15	2	40,00
50%	56	1	1	1	1	0	2500	58	0,43	0,42	11,0	14	72,9	14	2	33,33
0%	51	1	1	0	1	0	2500	17	0,60	0,33	0,0	0	233,2	17	14	77,33
0%	55	1	1	1	1	0	3500	33	0,33	1,18	0,0	0	195,6	29	3	32,67
0%	55	1	1	1	1	0	3500	33	0,33	1,18	0,0	0	187,3	23	3	32,67
0%	22	1	1	1	1	0	3500	33	0,33	1,18	0,0	0	230,9	27	3	32,67
0%	40	0	0	0	1	0	1000	25	0,54	-0,94	0,0	0	160,6	25	29	23,33
0%	39	1	1	1	1	1	3000	17	0,67	0,83	0,0	0	244,1	29	2	33,33
0%	35	1	1	0	1	0	2000	33	0,33	-1,34	0,0	0	418,3	24	3	72,00
0%	49	1	1	1	1	0	3500	58	0,23	-0,21	0,0	0	464,9	36	7	24,67
0%	45	1	1	1	1	0	2500	25	0,33	-2,14	0,0	0	239,7	20	3	59,33
0%	11	0	0	1	0	0	2500	25	0,33	-2,14	0,0	0	63,4	19	3	59,33

And lastly, the assumptions that slope and public transport attractiveness of a residential location play a role in the share of walking trips of survey respondents is also subordinate to the more decisive individual necessities: it may be true in some cases, but not statistically significant. For the Stuttgart region with its hilly topography, assumptions about the slope of residential locations actually lend more thought towards the popular local portrayal of the social strata, where more affluent parts of the population (and particularly the automobile industry) have historically build their residences in the hills, and workers had to live in the polluted valley in close proximity to the factory plants along the river Neckar. Now this picture may still hold merit for some reminiscent images of the region, but urban sprawl has been such that apart from the local green belt all modes of transport as well as demand-driven facilities are now servicing the region, regardless of topography. Public transport attractiveness is in the same category: it certainly plays a role in some people's mobility behaviour, but it can actually lead to less or more walking depending on the actual distance to the public transport stop. What is true, however, is an increasing demand for intermodal transport options, i.e., bike and ride facilities, bike transport opportunities on trains, hire bike systems in the inner city, etc. All of these options usually involve walking trips as well, on top of the physical activity that such trips require on their own. City planners are well advised to support and implement such opportunities if walking and physical activity is on their best practice agenda.

CONCLUSION

This chapter has started off with a comparison of mode shares of walking across a range of worldwide and European city regions. It has asked the question if environmental characteristics of local neigbourhoods are responsible for these differences, or if there might be more to it, mainly in the form of people's behavioural patterns. To find an answer, the author has taken a closer look at assessment methodologies for walkable communities, like walkability audits or GIS-based spatial indicators like the walkscore, the walkability index or a measure called the non-motorized car dependency for the Region of Stuttgart, Germany. Implementations of the latter three have been compared against a mobility database where survey respondents have given detailed personal and household characteristics and recorded their mobility activities over a period of at least two weeks. The comparison was conducted as regression analysis and has shown that no theoretical model can explain higher shares of walking within the research design. One result was that the walkability indicators underestimate the share of walking in the inner city, and overestimate that share for suburban zones. But apart from these results, quantitative analysis options have not yielded any results worthy of deeper consideration.

Not ruling out that more refined statistical analysis may be capable of detecting patterns in the data, an obvious conclusion is becoming more and more self-evident: travel behaviour in highly interconnected city regions is such that only few origins and destinations are within the assessed immediate neighbourhood, and personal predictors of mobility patterns are too diverse to be captured in simple household survey variables. Transport systems are highly interconnected, transport options are becoming increasingly diverse, and modern lifestyles are such that business related trip chains, dispersed social networks or even multiple places of residence add a level of complexity to the research that cannot be solved, at least not with a region-wide scope with almost 10.000 observations. Nevertheless, there is some inherent value in the employment of walkability indicators, even though they may not be fit to represent travel behaviour. City planners and public health authorities are still well-advised to use planning principles that focus on walkable neighbourhoods and a high level and quality of transport choices to provide the necessary options. Researchers, however, have to be careful with indicator interpretations, as to not pretend to represent reality without a sound scientific basis.

References

Bucksch, J. & Schneider, S. (Hrsg.) 2014. Walkability - Das Handbuch zur Bewegungsförderung in der Kommune, Bern: Verlag Hans Huber.

Bühler, R. & Jung, W. 2013. Transportation and Land-Use Planning in Germany and the U.S.: Lessons from the Stuttgart and Washingto, DC Regions, AICGS Policy Report. 53. Washington, DC: American Institute for Contemporary German Studies, John Hopkins University.

Djurhuus, S., Hansen, H. S., Aadahl, M. & Glümer, C. 2014. Individual Public Transportation Accessibility is Positively Associated with Self-Reported Active Commuting. Frontiers in Public Health, 2, S. 240.

Dobesova, Z. & Krivka, T. 2012. Walkability Index in the Urban Planning: A Case Study in Olomouc City. In: BURIAN, J. (Hrsg.) Advances in Spatial Planning. Rijeka: InTech, S. 179-197.

Ewing, R. & Cervero, R. 2010. Travel and the built environment. A meta-analysis. Journal of the American Planning Association, 76, S. 256-294.

Fina, S. 2013. Indikatoren der Raumentwicklung. Flächeninanspruchnahme und Landschaftszersiedelung. Dr. rer. nat., Eberhard Karls Universtität Tübingen. .

Frackelton, A., Grossman, A., Palinginis, E., Castrillon, F., Elango, V. & Guensler, R. 2013. Measuring walkability: Development of an automated sidewalk quality assessment tool. Suburban Sustainability, 1, S. 4.

Gärling, T. & Schuitema, G. 2007. Travel demand management targeting reduced private car use: effectiveness, public acceptability and political feasibility. Journal of Social Issues, 63, S. 139-153.

Giles-Corti, B., Macaulay, G., Middleton, N., Boruff, B., Bull, F., Butterworth, I., Badland, H., Mavoa, S., Roberts, R. & Christian, H. 2015. Developing a research and practice tool to measure walkability: a demonstration project. Health promotion journal of Australia, 25, S. 160-166.

Glazier, R. H., Creatore, M. I., Weyman, J. T., Fazli, G., Matheson, F. I., Gozdyra, P., Moineddin, R., Shriqui, V. K. & Booth, G. L. 2014. Density, destinations or both? A comparison of measures of walkability in relation to transportation behaviors, obesity and diabetes in Toronto, Canada. PloS one, 9, S. e85295.

Jack, E. & Mccormack, G. R. 2014. The associations between objectively-determined and self-reported urban form characteristics and neighborhood-based walking in adults. International journal of behavioral nutrition and physical activity, 11, S. 71.

Karusisi, N., Thomas, F., Méline, J., Brondeel, R. & Chaix, B. 2014. Environmental conditions around itineraries to destinations as correlates of walking for transportation among adults: the RECORD cohort study. PloS one, 9, S. e88929.

Lavoie, M., Burigusa, G., Maurice, P., Hamel, D. & Turmel, E. 2014. Active and safe transportation of elementary-school students: comparative analysis of the risks of injury associated with children travelling by car, walking and cycling between home and school. Chronic diseases and injuries in Canada, 34, S. 195-202.

Meyer, M. D. 1999. Demand management as an element of transportation policy: using carrots and sticks to influence travel behavior. Transportation Research Part A: Policy and Practice, 33, S. 575-599.

Region Stuttgart. Available: *http://www.region-stuttgart.de/* [Accessed 24 April 2015 2015].

Reyer, M., Fina, S., Siedentop, S. & Schlicht, W. 2014. Walkability is Only Part of the Story: Walking for Transportation in Stuttgart, Germany. International Journal of Environmental Research and Public Health, 11, S. 5849-5865.

Scanlin, K., Haardoerfer, R., Kegler, M. C. & Glanz, K. 2014. Development of a pedestrian audit tool to assess rural neighborhood walkability. Journal of physical activity & health, 11, S. 1085-1096.

Siedentop, S., Fina, S. & Krehl, A. 2014. Greenbelts in Germany's regional plans - an effective growth management policy? Landscape and Urban Planning, S.

Siedentop, S., Roos, S. & Fina, S. 2013. Ist die „Autoabhängigkeit" städtischer Siedlungsgebiete messbar? Entwicklung und Anwendung eines Indikatorenkonzepts in der Region Stuttgart. Raumforschung und Raumordnung, 71, S. 329–341.

Verband Region Stuttgart 2011. Begleituntersuchungen zur Fortschreibung des Regionalverkehrsplans - Band 1: Mobilität und Verkehr in der Region Stuttgart 2009/2010; Regionale Haushaltsbefragung zum Verkehrsverhalten, Schriftenreihe. 29.

Zwald, M. L., Hipp, J. A., Corseuil, M. W. & Dodson, E. A. 2014. Correlates of Walking for Transportation and Use of Public Transportation Among Adults in St Louis, Missouri, 2012. Preventing Chronic Disease, 11, S. E112.

In: Urban and Built Environments
Editor: Alexis Cohen
ISBN: 978-1-63483-067-6
© 2015 Nova Science Publishers, Inc.

Chapter 3

EVALUATION OF CHANGES IN URBAN ENVIRONMENTS USING ORTHOPHOTO MAPS

Marica Starešinič, Jure Ahtik, Andrej Učakar and Dejana Javoršek
University of Ljubljana, Faculty of Natural Science and Engineering, Ljubljana, Slovenia

ABSTRACT

Nowadays, urban environments are changing and expanding very fast, since people are moving to cities to get jobs, more opportunities and generally a better life. To gain control of this fast development and building projects, orthophotography images are used to display actual land and building features.

In our research, the aerial images, acquired by a model multicopter equipped with a digital camera, were analyzed and compared to the images of orthophoto maps available on selected web providers such as Google Earth, Geopedia, Apple Maps and Bing Maps.

Analyzed images were obtained/sampled from the selected area in Slovenia, in urban environment: The city of Ljubljana. Analysis was based on pedestrian areas, traffic areas, rooftops and green areas. In addition, acquired colors of Ljubljana (from the multicopter with a digital camera and orthophoto maps) were compared to the reference colors obtained from previous research.

The results of the analysis will show the differences between images acquired from the multicopter with a digital camera and images obtained from orthophoto maps. The research will also present advantages and disadvantages of using two different methods for obtaining images from the selected area of the city environment. Even more, comparison of colors that are typical for the selected city area will give the optimal method for the preservation of landscapes.

This kind of research is very important for sustainable development of urban environments to gain control of the development of construction sites to monitor the keeping of the building regulations and the use of the building materials (i.e., according to the construction regulations and the requirements for landscape preservation). Since the weather patterns are changing very fast, the results of this research will also give important insight on the environmental changes.

Keywords: orthophotography images, multicopter, spectroradiometry, digital photography, urban environment

1. Introduction

Aerial photography takes photographs of the ground from an elevated position. Balloons are probably the oldest platforms for aerial observation. As a matter of fact, aerial photographs of Paris were already captured in 1858 by Tournachon aboard a hot-air balloon [1]. Later on, and due to the simplification of the camera technology, other means such as kites (used by the English meteorologist E. D Archibald in 1882) and rockets (as used by the Swedish inventor Alfred Nobel in 1897) were utilized for aerial photography. Perhaps one of the most exciting early experiments was the use of small cameras mounted on the breasts of pigeons from the Bavarian Pigeon Corps, as proposed by J. Neubronner in 1903. Thus, Tournachon's adventures aside, one may conclude that the oldest form of aerial remote sensing was actually based on remotely piloted vehicles. Yet, the reader would be right to harbor doubts about defining a pigeon as an ''unmanned platform'' or ''remotely-piloted aircraft''.

The use of unmanned aerial vehicles for civil and commercial purposes has been rising in recent years. Similarly, a number of new applications have become possible in the area of photography, geodesy, particularly in photogrammetry, infrastructure monitoring [2] and remote sensing [3]. Platforms for aerial photography include fixed-wing aircraft, helicopters, multirotor Unmanned Aircraft Systems (UAS), balloons, blimps and dirigibles, rockets, pigeons, kites, parachutes, stand-alone telescoping and vehicle mounted poles. Mounted cameras may be triggered remotely or automatically [4].

A multicopter is a mechanically simple aerial vehicle whose motion is controlled by speeding or slowing multiple downward thrusting motor/ propeller units. Multicopters are aerodynamically unstable and absolutely require an on-board computer (flight controller) for a stable flight. The flight controller combines data from small on-board MEMs gyroscopes, accelerometers (the same as those found in smart phones) to maintain an accurate estimate of its orientation and position [5].

The technology has developed, and now we can see the terrain from above [6] with airplanes, helicopters and also from space using satellites.

The colors of urban environments [7] were analyzed by researchers using digital photography as described in the paper by A. Gou [8] where they defined the colors in urban environment in China to provide dominant colors for urban planning, guides for architects and usage of new materials and technology. The colors of stonewalls in England [9] were analyzed to improve restoration work. Digital cameras are capable of measuring small color differences as it was evaluated in the research, where samples were compared with the spectroradiometer [10].

Scenes viewed in different daylight conditions were analyzed by Romero [11]. His intention was to determine the constant sensor response ratio in different daylights. Kwon also discovered that the best camera settings are to be determined by experimental work on selected illumination [12].

The majority of the world's population now resides in urban environments and information on the internal composition and dynamics of these environments is essential to enable preservation of certain standards of living [13]. The study, how people shape their environments, was conducted by Rapoport [14] and it is presented that the people interact differently with their urban environment based on landscape, colors and also on the interactions in society. How to plan the autonomous vehicle navigating through urban environments, presented by Ferguson [15], is based on the solutions for problems both on-road and also on unstructured areas of the environment.

In fact, urban surfaces can represent more challenge than many other types of environments, because of the very large variety of materials concentrated in relatively small areas, as presented by Mereny [16], maps can give critical support for accurate identification of surface objects and conditions, but a detailed and precise segmentation of remote sensing spectral images remains a much needed key component in the comprehensive analysis and understanding of surfaces.

Computer vision is also used for reconstruction of urban environments [17] presented system collects video, GPS and INS data, which are processed in real-time to produce geo-registered, detailed 3D models that represent the geometry and appearance of the world.

Such information can be used to model cities from unstructured point data [18] where they use geometric primitives to model cities, and they have experimentally validated the approach on complex urban structures and large urban scenes of millions of points.

2. EXPERIMENTAL

In our research, aerial photos were taken from video acquired by an octocopter type X-8 [19]. The octocopter consists of 8 contrary rotating motors, which provide flight time and enough power for a lifting camera with an electronically stabilized camera gimbal. Images were recorded at height of 465m above the Ljubljana Castle. Camera that was used was Sony HDR CX730E [20] using optical stabilization system Carl Zeiss® Vario-Sonnar T* lens. The mentioned camera has integrated 1/2.88 type (6.3mm) back-illuminated Exmor R™ CMOS Sensor 6140K pixels in format 16:9. The clip was recorded using MPEG4-AVC/H.264 AVCHD™ movie format, HD: 1920x1080/50p (PS) resolution and HD PS: Approx.28Mbps/FX Average Bit Rate. The octocopter has an integrated 3 axial camera gimbal system which provides electronically stabilized pictures and therefore stabilization in the post production using algorithms in programs such as Adobe after Effects [21] is not necessary.

Aerial images, acquired by a model octocopter equipped with a digital camera, were analyzed and compared to the images of orthophoto maps available on selected web providers such as Google Earth [22] (Google), Geopedia [23], Apple Maps [24] (Apple) and Bing Maps [25] (Microsoft). All images were adjusted to the same size so that they cover the same area of urban environment, i.e., the city of Ljubljana (Figure 1) between the following coordinates: 46 03'06.2"N 14 30'17.6"E, 46 03'04.8"N 14 30'26.7"E, 4602'52.6"N 14°30'12.5"E, 46°02'51.6"N 14°30'24.0"E. White balance on all images has been set to the point 46°02'54.5"N 14°30'19.7"E with Adobe Lightroom 5. Using Adobe Photoshop CS6 images were auto aligned, so that the same areas became overlapped. Separation of roads, roofs, water and green areas was done using Google vector maps. The percentages of each area are shown in Table 1. All the colors in the separated area of each map have been averaged with Adobe Photoshop CS6 (Avarage blur).

Table 1. The percentages of each area

area	area percentages (%)
street	11.3
buildings	41.9
water	39.3
green	7.4

Figure 1. Image of the city of Ljubljana (image was taken from video).

In addition, colors of Ljubljana obtained in our research from a multicopter with a digital camera and orthophoto maps were compared to the reference colors from previous researches, where street area of old historic town was measured with spectroradiometer PR-650 (X-Rite) with measurement range 380-780nmon distance 50 m [26], CIEL*a*b* values were calculated using CIE 2 color-matching functions and calculated tristimulus values for illuminant that could be compared to standard illuminant D65 with coordinates $X_0 = 5.60$, $Y_0 = 100.00$, $Z_0 = 107.51$. Color differences between average CIEL*a*b* values calculated from RGB values and measured average CIEL*a*b* values were calculated using equation ΔE^*ab.

3. Results

Urban environments were divided on pedestrian areas, traffic areas, rooftops and green areas, as seen in the video. Visualization of roads, buildings, water and green areas on the images (video), obtained from a multicopter with a digital camera are presented in Figure 2.

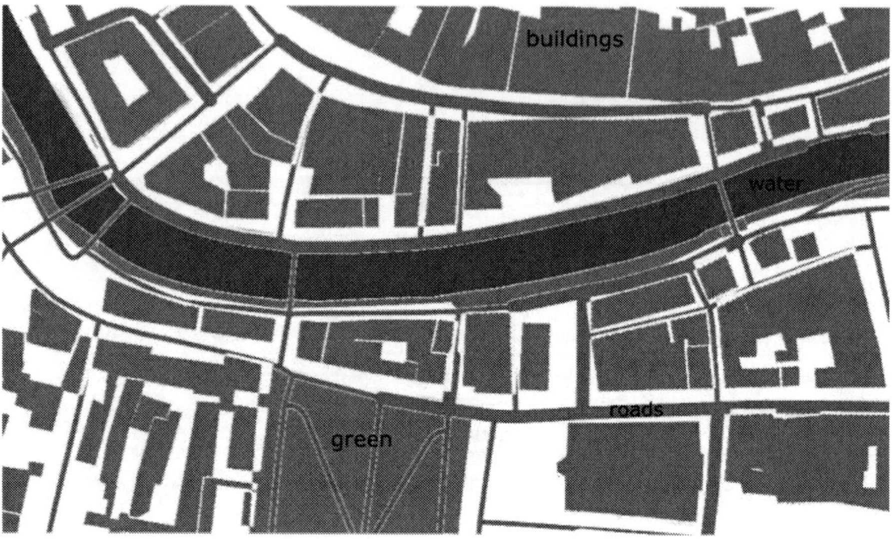

Figure 2. Visualization of roads, buildings, water and green areas on the images (video).

Figure 3 shows the results of RGB values obtained from images acquired from a multicopter with a digital camera. Those RGB values were compared to the images of orthophoto maps available on selected web providers such as Google Earth (Google), Geopedia, Apple Maps (Apple) and Bing Maps (Microsoft). Results of calculated average differences in red (dR), green (dG) and blue (dB) component as well as average color differences dRGB are presented in Figures 4-7.

Figure 4 shows the results of the traffic area with grey component where the largest average color differences dRGB are obtained with Google, followed by Apple and Microsoft, while the Geopedia shows the best results. In the case of Apple and Google, all three RGB components of color are more saturated than the RGB values obtained from the Video, while in the case of

Microsoft and Geopedia, blue (B) component is reduced, as indicate negative values of average dB.

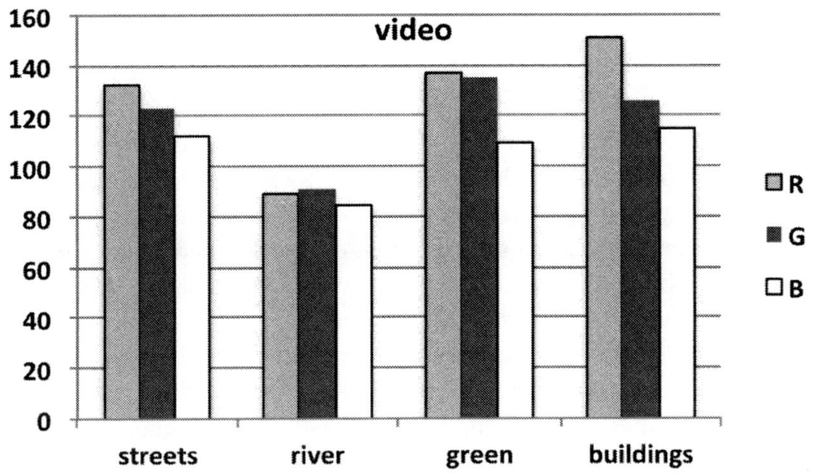

Figure 3. RGB values obtained from images acquired from multicopter.

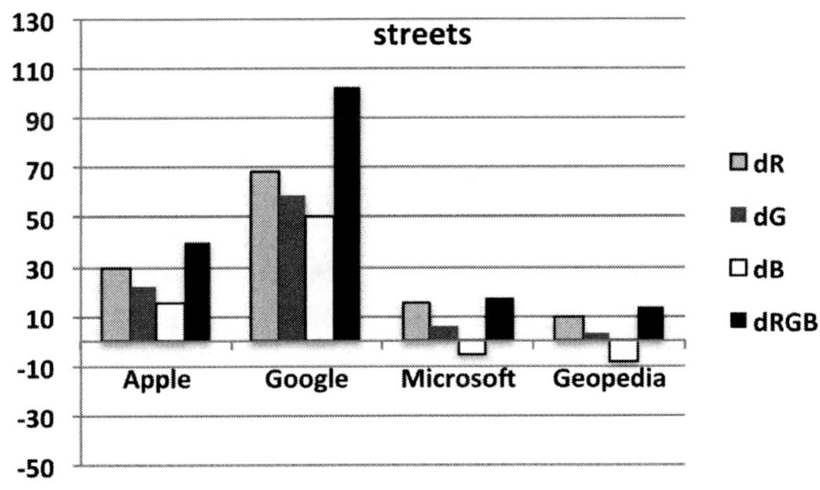

Figure 4. Average color differences dRGB, dR, dG and dB between RGB values acquired from multicopter and RGB values from different providers in the traffic area (streets).

Figure 5 shows the results of the water area (river Ljubljanica) where the largest average color differences are obtained with Google, followed by Apple and Geopedia with the same dRGB, while Microsoft shows the best results. In the case of Apple, Google and Geopedia, all three RGB components of color are more saturated than the RGB values obtained from the video, while in the case of Microsoft, blue (B) component is reduced, as indicated by negative values of an average dB.

Figure 6 shows the results of the green area where the largest color differences are obtained with Google, followed by Apple and Microsoft, while Geopedia shows the best results with the largest negative value of average dB.

Figure 7 shows the results of the rooftops area where the largest color differences are obtained with Google, followed by Apple and Microsoft, while Geopedia shows the best results.

In comparison with the results of different areas of the Ljubljana city obtained from Video, the largest average color differences occurs in the case of color of the rooftops - red color and the smallest one occurs in the case of color water area - dark green.

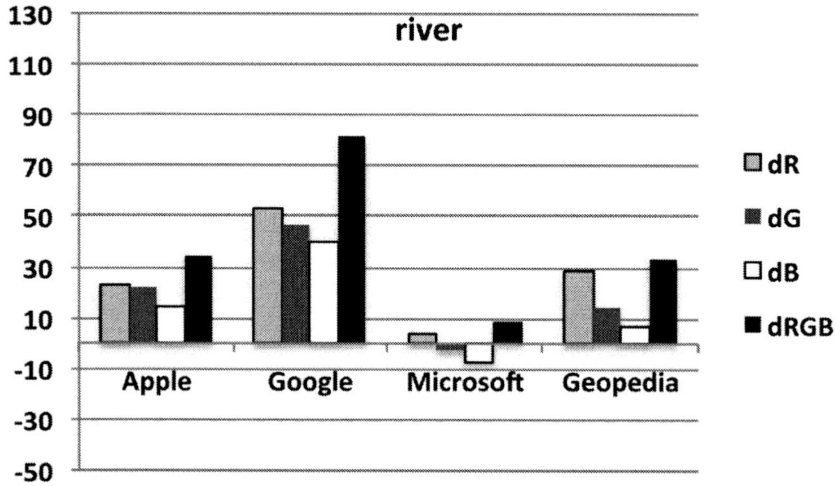

Figure 5. Average color differences dRGB, dR, dG and dB between RGB values acquired from a multicopter and RGB values from different providers in the water area (river).

Our results were also compared with measurements made with spectroradiometer PR-650 [26]. Results of average color difference between average CIEL*a*b* values calculated from RGB values (images of traffic area) and measured average CIEL*a*b* values are shown in Table 2.

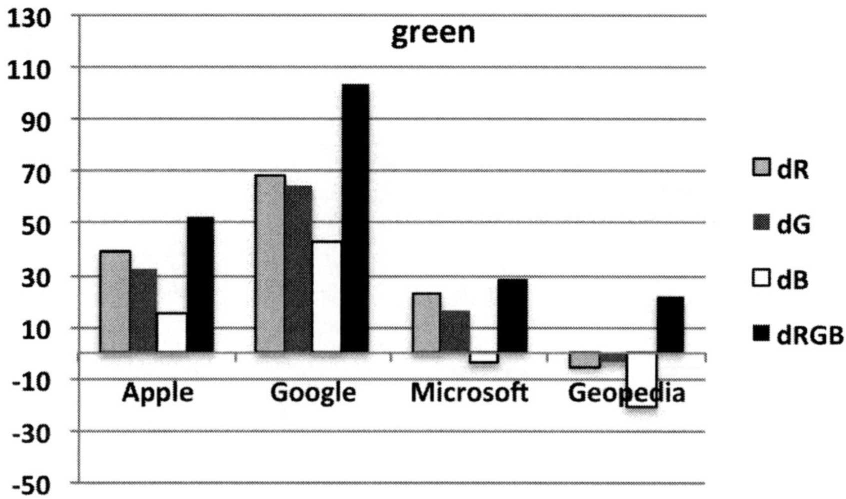

Figure 6. Average color differences dRGB, dR, dG and dB between RGB values acquired from a multicopter and RGB values from different providers in the green area (green).

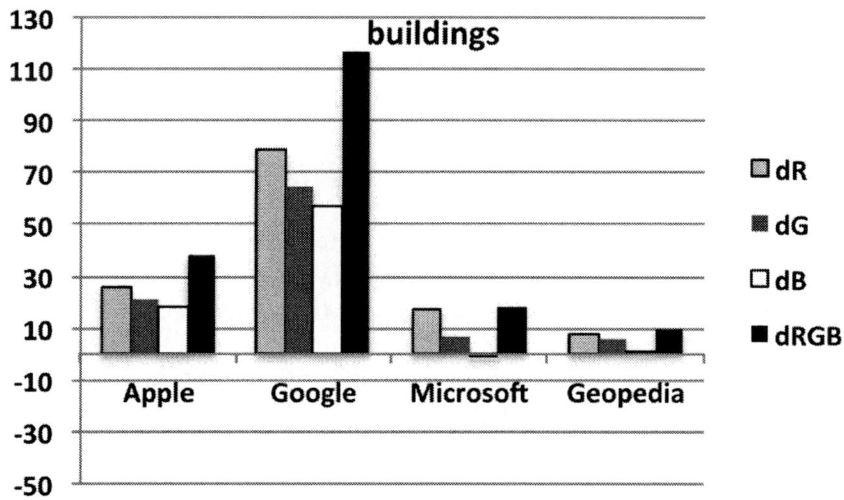

Figure 7. Average color differences dRGB, dR, dG and dB between RGB values acquired from a multicopter and RGB values from different providers in the rooftops area (buildings).

Table 2. Color differences between average CIEL*a*b* values calculated from RGB (images of traffic area) and measured average CIEL*a*b* values

Provider	R	G	B	L^*	a^*	b^*	ΔE^*ab
Apple	102	101	97	42.30	0.49	1.80	3.67
Google	64	64	62	27.03	-0.45	1.25	12.11
Microsoft	117	117	117	49.26	0.00	0.00	4.83
Geopedia	122	120	120	50.62	0.68	0.24	4.64
Video	132	123	112	51.95	1.81	7.05	2.43
PR650				48.15	-2.27	4.80	

Results of calculated color differences show perceivable differences for images obtained from Apple, Microsoft, Geopedia and Video. Exceptions are the images obtained from Google, since the images are fundamentally much darker ($\Delta L^* = 21.12$). Minimum color difference ΔE^*ab (2.43) was obtained when the picture was taken from the video, which is in the category of small color differences [27].

Based on the results of our research, we can conclude that for evaluation of some general changes in urban environments orthophoto maps could be used, while for colorimetric accuracy the Video should be used, since the images had the minimum color difference, while other images obtained from Apple, Microsoft and Geopedia show perceivable differences.

CONCLUSION

In the first part of our research, we compared areal video images acquired by a model octocopter, to the images acquired by Google Earth (Google), Geopedia, Apple Maps (Apple) and Bing Maps (Microsoft), and we can conclude, that the images with the smallest dRGB were obtained from Microsoft Bing Maps and Geopedia, followed by Apple Maps, while images from Google Earth were much darker and saturated.

In the comparison of images taken with all five media to the measurements made with the spectroradiometer PR-650 from former research, we can conclude that the results are surprisingly good and color difference ΔE^*ab is acceptable in most cases except in the data from Google.

From our research of the evaluation of changes in urban environments using orthophoto maps, we can conclude that for the first impression of our environments, we can use images from web providers as they give surprisingly good results, but for the more detailed research and planning, the look from above, with a video camera, would still be required.

Since the weather patterns are changing very fast, the results of this research will also give important insight on the environmental changes.

REFERENCES

[1] Colomina, I., Molina, P., Unmanned aerial systems for photogrammetry and remote sensing: A review, *ISPRS Journal of Photogrammetry and Remote Sensing* 92 (2014) 79-97, cited 20 Jan 2015, (e-text type) http://www.sciencedirect.com/science/article/pii/S0924271614000501.

[2] Oblak, Tomaž, *Uporaba brezpilotnih letal v Sloveniji:* Graduation thesis, cited: 12 Jan 2015, (e-text type) https://dk.um.si/Dokument.php?id=6858.

[3] Zajc, Sebastjan, *Uporaba brezpilotnih plovil pri množičnem urejanju parcelnih mej: Graduation thesis, cited:* 10 Jan 2015, (e-text type) http://drugg.fgg.uni-lj.si/4902/1/BGG057_Zajc.pdf.

[4] Wikipedia, the free encyclopedia, *Aerial photography,* last modified and cited on 25 January 2015, (e-text type) http://en.wikipedia.org/wiki/Aerial_photography#Early_history.

[5] *APM Autopilot site: What is a MultiCopter and How Does it Work:* cited: 20 jan 2015, (e-text type) http://copter.ardupilot.com/wiki/what-is-a-multicopter-and-how-does-it-work/.

[6] Arthus-Bertrand Yann, *Earth from above*: < http://justpaste.it/3ky > ali tudi knjiga Yann Arthus-Bertrand, Earth from above, Publisher: Harry N. Abrams; 3rd edition (2005).

[7] Starešinič, Marica, Simončič, Barbara, Bračko, Sabina: Using a digital camera to identify colors in urban environments. *Journal of imaging science and technology*, ISSN 1062-3701, 2011, vol. 55, no. 6 (2011).

[8] Gou Aiping and Wang Jiangbo: *Research on the Location Characters of Urban Color Plan in China, Color Research and Application,* vol. 33, no. 1, pp. 68-76, (2008).

[9] Thornbush, M. J., Grayscale calibration of Outdoor Photographic Surveys of Historical Stone Walls in Oxford, England, *Color Research and Application*, vol. 33, no. 1, pp. 61-67, (2008).

[10] Valencia E., Millan M. S., Small Color Differences in the Very Pale and Dark Grayish Regions Measured by Camera, *Journal of Imaging Science and Technology,* vol. 49, No.6, pp. 605-619. (2005).

[11] Romero, J. et al., Sensor-response-ratio constancy under changes in natural and artificial illuminants, *Color Research & Application,* vol. 32, No. 4, pp. 284-292, (2007).

[12] Kwon O., Cho Y., Ha Y., Illumination Estimation Based on Valid Pixel selection from CCd Camera Response, *Journal of Imaging Science and Technology,* vol. 48, No.3, pp. 040906-040906-11, (2005).

[13] Phinn S., Stanford M., Scarth P., Murray A. T. & Shyy P. T., Monitoring the composition of urban environments based on the vegetation-impervious surface-soil (VIS) model by subpixel analysis techniques, *International Journal of Remote Sensing,* Volume 23, Issue 20, (2002).

[14] Rapoport, Amos: *Human Aspects of Urban Form: Towards a Man: Environment Approach to Urban Form and Design*. Elsevier, (2013).

[15] Ferguson, Dave, Thomas M. Howard, and Maxim Likhachev. Motion planning in urban environments. *Journal of Field Robotics*, 939-960 (2008).

[16] Merényi, Erzsébet, Beáta Csathó, and Kadim Tasdemir:Knowledge discovery in urban environments from fused multi-dimensional imagery,*Urban Remote Sensing Joint Event, 2007*. IEEE, (2007).

[17] Mordohai, Philippos, et al., *Real-time video-based reconstruction of urban environments,* ISPRS Working Group 4 (2007).

[18] Lafarge, Florent, and Clement Mallet, Building large urban environments from unstructured point data. Computer Vision (ICCV), 2011 *IEEE International Conference* on. IEEE, (2011).

[19] APM Autopilot site: *Advanced MultiCopter Design:* cited: 20 jan 2015, available (e-text type) http://copter.ardupilot.com/wiki/advanced-multicopter-design/.

[20] Sony online technical support web page: *Sony HDRCX730E Technical specification,* cited: 18 jan 2015, (e-text type) http://www.sony.co.uk/support/en/content/cnt-specs/HDR-CX730E/list.

[21] Adobe.com. *Learn and support web page: Adobe after Effects: How to stabilize shaky video footage,* cited: 10 jan 2015, (e-text type) https://helpx.adobe.com/after-effects/how-to/stabilize-footage.html.

[22] *Google Earth,* (e-text type) https://www.google.com/earth/.

[23] Geopedia, si, *interaktini atlas Slovenie,* Sinergise, (e-text type) http://www.geopedia.si/#T105_x499072_y112072_s9_b4.

[24] *Apple Maps Connect,* (e-text type) https://mapsconnect.apple.com.

[25] *The Bing Maps Platform,* (e-text type) http://www.microsoft.com/maps/.

[26] Starešinič, Marica, Ahtik, Jure, Simončič, Barbara, Bračko, Sabina:Determination of object colours in urban environments using spectroradiometry. *Tekstilec*, ISSN 0351-3386, vol. 55, No. 2, pp. 87-91, (2012).

[27] Schläpfer, K., Farbmetrik in der grafishen Industrie, Dritte Auflage. *UGRA,* 61 (2002).

In: Urban and Built Environments
Editor: Alexis Cohen
ISBN: 978-1-63483-067-6
© 2015 Nova Science Publishers, Inc.

Chapter 4

COMPREHENSIVE ANALYSIS OF VOC EMISSION SOURCES IN BELGRADE URBAN AREA

A. Stojić[1], S. Stanišić Stojić[2], Z. Mijić[1], L. Ilić[1], M. Tomašević[1], M. Todorović[1] and M. Perišić[1]
[1]Institute of Physics Belgrade, University of Belgrade, Belgrade, Serbia
[2]Singidunum University, Belgrade, Serbia

ABSTRACT

The identification of VOC emission sources and analysis of their impact on air quality in complex Belgrade (Serbia) urban environment is presented. The concentrations of VOC were measured by using Proton Transfer Reaction Mass Spectrometer, whereas the concentrations of other pollutants (NO_x, NO, NO_2, SO_2, CO, and PM_{10}) and meteorological parameters were obtained from the automatic monitoring station located at the same site. Two US EPA receptor models, Positive Matrix Factorization and Unmix, were applied on the obtained dataset, and the resolved profiles were analyzed and compared. The divergent results suggesting weaknesses in used models are discussed in details. In addition to this, statistical analyses including bivariate polar plot, bivariate cluster and conditional bivariate probability function were performed within the statistical software environment R using the Openair package. The results enabled reliable identification of the emission sources with regards to several factors, such as: pronounced daily and weekly variations, day-of-week difference, statistical indicators, relationships with meteorological data - particularly wind speed and direction, toluene to benzene ratios, *etc.* Apart from the local sources attributed to vehicle exhaust, aged plumes, solid fuel burning and local laboratory, significant contributions of industrial facilities, including oil refinery, petrochemical complex, and chemical industry located in the suburban area of Belgrade were observed.

INTRODUCTION

There has been a long-standing interest in volatile organic compounds (VOC) based on their detrimental effects on human health and the environment. They represent a heterogeneous group of reactive trace species originating from a variety of biogenic and anthropogenic sources. Depending on the level and nature of the compound, both short- and long-term exposure to VOC can be associated with a wide range of adverse health effects (Huang et al., 2011). Thereby, several compounds are known or suspected carcinogens, while some of them are capable of inducing vital organ disorders (Jung et al., 2012, Romagnoli et al., 2014). As previously reported, the population in highly industrialized areas, belonging to lower socioeconomic groups, as well as young and elderly people, appears to be more susceptible to pollution-related mortality effects (Richardson et al., 2013).

Apart from the impact on human health, VOC are involved in several environmental issues associated with the climate change and increase in the atmospheric oxidizing ability. Furthermore, numerous studies have underlined their importance in biogeochemical cycles, the formation of secondary organic aerosols, greenhouse gases and other damaging pollutants, such as hydroxyl radical ($OH^{\bullet}$), nitrate radical ($NO_3^{\bullet}$), and ozone (O_3), which have the potential to affect the radiative energy balance of the Earth-troposphere system (Barley et al., 2011, Ran et al., 2011, Huber and Knutti, 2008).

Biogenic emissions from forests, oceans and soils consisting mainly of isoprene (C_5H_8), monoterpenes and associated oxygenated species have been acknowledged as the major VOC source on a global scale so far (Kansal, 2009). The most abundant are plant volatiles resulting from constitutive and induced emissions, out of which the latter increase in response to environmental stress. Their ecological role is to act as plant-plant and plant-animal interaction signals, and serve as protection against herbivore attacks, high temperature, high irradiance and oxidative stress (Hare, 2011, Rodriguez-Saona et al., 2013, Seco et al., 2011).

The key sources estimated to be the major VOC pollution contributors in urban areas are anthropogenic, e.g.: traffic in Beijing, China and Seoul, South Korea (Na et al., 2005, Song et al., 2007); petrochemical industry in Izmir, Turkey (Elbir et al., 2007); solvent usage at San-Diego-Tijuana border, Mexico (Zheng et al., 2013); etc. The most abundant species are monoaromatic hydrocarbons, commonly known as BTEX (benzene, toluene, ethylbenzene, and xylenes), which are used as indicators of man-made pollution in populated areas. Due to their carcinogenic and neurotoxic effects, BTEX are classified as priority pollutants by the Environmental Protection Agency (U.S. EPA).

Depending on their atmospheric lifetimes and meteorological conditions, locally generated VOC are considered to be of influence in remote, downwind locations at spatial scales of several hundred kilometers and temporal scales on the order of several days (Brankov et al., 2003). Therefore, research aimed at identification of VOC sources and estimation of their contributions is crucial for design and implementation of efficient environmental regulations and control strategies. In the previous studies, the multivariate receptor models, such as Positive Matrix Factorization (PMF) (Brown et al., 2007), Unmix (Olson et al., 2007) and Chemical Mass Balance (Song et al., 2008), have been widely used for the purpose of VOC source attribution and profiling. The identification of the sources has been based mainly on the chemical composition of the profile, in spite of a few major limitations of such approach, which can seriously affect the reliability of the conclusions. Firstly, the profiles are resolved under the assumption that statistically correlated species share the same source. Nevertheless, significant variations in VOC atmospheric residence time, observed as a result of the loss of highly reactive species through photochemical and deposition processes, can disable proper profile separation and increase discrepancy between the model solution and reality (Belis et al., 2013, Yuan et al., 2012). Thus, using more than one model is highly recommended, and according to Henry and Christensen (2010), it can also minimize the impact of dataset quality on model performance. Another issue is associated with the discrimination capability of Proton Transfer Reaction Mass Spectrometer (PTR-MS), which is limited in terms that each registered mass may relate to more than one compound due to potential mass interferences of isobaric species. Additionally, the formation of cluster ions and fragmentation of product ions may affect the interpretation of the mass spectra. Finally, it should be mentioned that most of the target species are emitted from a number of sources, while only few of them are considered unique tracers. Therefore, additional analysis is required to enable reliable source identification and constrain the uncertainties in the source contribution estimates.

This chapter presents a comprehensive analysis of VOC pollution in Belgrade urban area. Measurements were conducted by means of PTR-MS, capable of simultaneous detection of several VOC with low detection limits of 30 ppt. Slow response methods have been used for identification of seasonal and diurnal patterns of the urban VOC concentrations. However, due to its high temporal resolution, PTR-MS is more suitable for detection of short-term variability. Two receptor models, PMF and Unmix, were applied for source apportionment, while the identification of the resolved profiles was conducted considering chemical compositions, daily and weekly variations, relationships with meteorological data, toluene to benzene ratios, etc. The episode of the extreme NE wind blowing up to 20.6 m s^{-1} from January 30th to February 2nd was registered and its impact on the observed mixing ratios was discussed. The conditional bivariate probability function (CBPF) was used to identify wind direction and wind speed intervals that are dominated by high pollutant concentrations. The contribution of regional and long-range transport was estimated using trajectory sector analysis (TSA), whereas potential source contribution function (PSCF) and concentration weighted trajectory (CWT) models were applied for the distribution of non-local potential source regions. The variations of planetary boundary layer (PBL) height were taken into account for all transport-related models, since its rise leads to an increase in the portion of lower atmosphere where the intense exchange of gases occurs, thus causing the enhanced transport of pollutants from remote areas (Watson et al., 2001).

In conclusion to this, the herein presented in-depth analysis enabled reliable identification of the local and remote VOC emission sources, and estimation of their respective contributions. Additionally, the factors influencing VOC temporal dynamics in Belgrade urban area were considered and discussed.

STUDY AREA

Belgrade is the capital of Serbia and an important regional traffic hub, with a population of 1.6 million of residents. Its industrial suburban area is home to a network of coal-fired power plants, petrochemical complex, chemical industry, manufactory of polymers and plastic resins and oil refinery. Apart from large industrial facilities, several factors associated with heavy reliance on coal in power production; outdated technologies, inefficiency and poor maintenance in all economic sectors, have a detrimental impact on air quality in the urban zone. In recent years, greater attention has been devoted to establishing balance between economic objectives and sustainable development. However, the environmental policy based on end-of-pipe solutions and failure to implement current laws and regulations still poses a major barrier to air quality improvement.

Besides VOC, particulate matter (PM) is also among major primary anthropogenic pollutants in Belgrade urban area (Tasić et al., 2006; Rajšić et al., 2008). PM is a complex mixture of organic and inorganic substances, including elemental and organic carbon, metals, nitrates and sulphates. Among the metals constituting the PM, of particular concern are As, Cd, Cr, Ni, and Pb, which are human carcinogens, thus determination of metals in atmospheric particles is of great interest. During the past decade several studies have been conducting dealing with source apportionment (Mijić et al., 2010, 2012; Tomašević et al., 2013a, Stojić et al., 2015a) and PM source emission reduction required in Belgrade (Mijić et al., 2009, Perišić et al., 2014). In addition to instrumental measurements, plant biomonitoring have also been applied to study the particle-bound trace element atmospheric contamination. Some plant species have shown to be effective biomonitors. Mosses have been widely considered among the most appropriate ones. Several naturally growing moss species have been studied to assess trace element atmospheric deposition (Aničić et al., 2007).

A series of studies were done with active moss biomonitors, i.e., moss transplanted in the so-called "moss bags," in search for correlation with atmospheric bulk deposition (Aničić et al., 2009). Active moss biomonitoring was also applied in studying of small-scale spatial distribution of airborne major and trace elements in the Belgrade urban area (Vuković et al., 2013). Leaves of some common urban tree species, widely known as accumulative biomonitors, were also studied for their biomonitoring features. Different species vary in their morpho-anatomical and physiological characteristics (Tomašević et al., 2005), and so are their abilities to either capture airborne PM deposited on leaf surfaces or to accumulate contaminants in their tissues, and to monitor environmental contamination, including spatial, seasonal, and temporal variations (Tomašević et al., 2008, Aničić et al., 2011). Studying the natural isotopes of lead, an attempt was made to use tree leaves as biomonitors for identification of the Pb sources in the urban air (Tomašević et al., 2013b).

The climate in Belgrade is moderate continental with all four seasons. Autumn lasts longer than spring, with longer sunny and warm periods, while winter is moderate, with an average of 21 days when the temperature is below 0°C. The mean annual temperature is 11.7°C, with the coldest and hottest months being January and July with 0.1°C and 22.1°C, respectively. During the winter and autumn months, a strong southeast wind Koshava, blows in 2-3 days intervals. Moreover, Koshava in Belgrade region has an average speed of 25-43 km h^{-1}, while certain strokes can reach up to 130 km h^{-1}. The mean annual volume of rainfall in Belgrade amounts to 669.5 mm, with the average annual insolation of 2,096 hours. The rainiest months are May and June. The average number of snowy days is 27, while an average thickness of snow cover varies between 14 and 25 cm. The mean atmospheric pressure in Belgrade is 1001 mbar, while mean relative humidity is 69.5%.

The measurement site was located at the Institute of Public Health in Belgrade (44°49' N, 20°28' E), in the residential, commercial and services area. The stagnant conditions within the urban canyon avenue with severely congested traffic flow and low ventilation led to inefficient pollutant removal and relatively high levels of oxygenated species. As in many other urban areas, traffic emissions are an important source of pollution, due to rapidly growing number of in-use vehicles, most of which are older than 10 years (NSO, 2011).

In Table 1, the basic statistics for VOC-related masses, NO_x, NO_2, NO, PM_{10}, SO_2, CO, and meteorological parameters are presented. When compared to the data from previously conducted studies, VOC levels are similar to or lower than those registered in the large industrial centers or overpopulated metropolis of Asia and Africa, but still higher than the levels measured in most of the major cities in Europe and the USA.

In the study examining personal exposure to VOC concentrations in several European cities, Saarela et al., (2003) reported that the most abundant single aromatic compound was toluene, accounting for 35-50% of the total sum of target aromatic species.

Additionally, neighboring Southern European cities, Milan and Athens, are shown to be the most polluted, whereas the pollutant concentrations decrease gradually towards the North and West. Thereby, it should be noted that diverse approaches to data presentation, as well as the effects of regionally specific conditions such as those related to meteorology, often complicate the comparison of reported values and need to be taken into consideration when evaluating regional budget.

METHODOLOGY

All measurements were conducted from January 22nd to March 24th, 2014 at the Institute of Public Health in Belgrade as part of the winter observation period (Figure 1, left). The concentrations of inorganic gaseous oxides and PM_{10} together with meteorological data were obtained from the automatic monitoring station. Standard methods for the measurement of the concentrations of: nitrogen dioxide and nitrogen monoxide by chemiluminescence (standard EN 14211:2005, device Horiba APNA 360), sulphur dioxide by ultraviolet fluorescence (EN 14212:2005, Horiba APSA 360), carbon monoxide by nondispersive infrared spectroscopy (EN 14626:2005, Horiba APMA 360), xylenes and ethylbenzene – pumped sampling followed by thermal desorption and gas chromatography (EN 14662-3, Chromatotec BTX), mass of particulate matter (PM_{10}) on a filter medium by beta-ray absorption method (ISO 10473, Thermo FH 62-IR) were used. Meteorological data (atmospheric pressure, temperature, humidity, precipitation, wind speed and direction) were obtained by Lufft WS500-UMB Smart Weather Sensor.

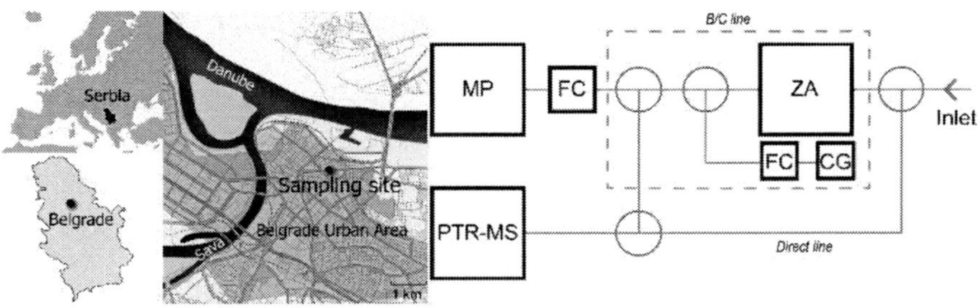

Figure 1. Measurement site (left) and experimental setup: Standard PTR-quad-MS, FC – flow controllers, ZA – zero air generator, CG – calibration gas, and MP – membrane pump (right).

On-line measurements of 36 VOC-related masses were conducted using PTR-MS (Standard PTR-quad-MS, Ionicon Analytik, GmbH, Austria). The details of the method have already been given elsewhere (Lindinger et al., 1998, de Gouw and Warneke, 2007). Experimental setup consisted of direct line used for VOC, and the line used for background and calibration (B/C) measurements (Figure 1, right). The inlet of the instrument, 3 m heated (70°C) silcosteel line with inner diameter 3 mm, was placed 3 m above ground. VOC-related masses with 0.5 s dwell time, and five control parameters (m/z 21, m/z

25, m/z 30, m/z 32, and m/z 37) data were obtained in 24 s cycles. Drift tube parameters included pressure in the range 2.08 to 2.11 mbar, temperature 60^0C, voltage 600 V, E/N parameter 145 Td, and reaction time 90 μs. The count rate of $H_3O^+H_2O$ was 1 to 7% of the average $5.1 \cdot 10^6$ counts s^{-1} count rate of primary H_3O^+ ions.

The calibration was done according to Taipale et al. (2008) using TO-15 Supelco gas mixture (m/z 57, m/z 79, m/z 93, m/z 107, and m/z 121) diluted with zero air (HC free, Messer) to five concentrations ranging from 0.5 to 100 ppb by means of ASGU 370-p HORIBA system. Normalized sensitivities were obtained in the range 6.2 to 14.3 npcs ppb^{-1}. Detection limit of 1-h averaged VOC were less than 0.5 ppb, except for methanol (2.0 ppb) and acetone (1.1 ppb). Key m/z-signals in the dataset were observed for 29 out of 36 masses (Galbally et al., 2008).

Two advanced multivariate receptor models, Unmix (USEPA, 2007) and Positive Matrix Factorization (Version 3.0) (USEPA, 2008), were applied to the 1169 1-hour-averaged concentrations of 29 species for the purpose of source apportionment (Table 1). The detailed descriptions of the models are presented elsewhere (Henry, 2003; Paatero, 1999). In this chapter the robust mode was used to handle outlier values with threshold distance $\alpha=4$. For concentrations below the detection limit (DL), the values were set to DL/2. The combination of trial-and-error was used to maximize the number of species included as input variables with the general goal to maximize number of the variables producing feasible and physically interpretable solution (Stojić et al., 2015). In order to evaluate uncertainty of the resolved profiles, 100 bootstrap runs with R^2-value of 0.6 were performed.

Receptor model results were evaluated using statistical tests (Fraction of predictions within a factor or two - FAC2, Normalized mean bias - NMB, Normalized mean gross error - NMGE, Root mean squared error - RMSE, Correlation coefficient - r, Coefficient of Efficiency - COE and Index of Agreement - IOA) (Wilks, 2011) and Taylor diagrams (Taylor, 2001) with the statistical software environment R (Team, 2012) using the Openair package (Carslaw and Ropkins, 2012).

Statistical analyses including bivariate polar plot, bivariate cluster (k-means clustering, grouping similar conditions together) (Carslaw and Beevers, 2013), and conditional bivariate probability function (CBPF) (Uria-Tellaetxe and Carslaw, 2014) were performed with the Openair package.

In order to differentiate between local emissions and transport contributions, a two-step procedure was introduced. The local contribution was excluded from the time series, thus providing a baseline by the use of frequency differentiated non-linear digital filtering algorithm implemented in the function baseline.rollingBall (wm=ws=6) of the Baseline package (Kneen and Annegarn, 1996) of the statistical software environment R. Subsequently, in order to distinguish between local background and regional transport contributions, TSA was applied to the resulting baseline. The area was divided into eight 45°-sectors (N, NE, E, SE, S, SW, W, and NW), and the contribution of the least-polluted sector was taken as background. The contributions of transport and local background to the total observed pollution were determined according to Zhu et al. (2011). Furthermore, the impact of non-local source regions and pollution transport pathways were investigated by the use of PSCF and CWT (Hsu et al., 2003). Air-mass back trajectories were computed using the NOAA Air Resources Laboratory HYSPLIT model (Draxler and Rolph, 2014). Daily 72-hour back trajectories, starting from the sampling site every hour UTC each day, were evaluated for six different heights above ground level (350, 500, 750, 1000, 1500, and 2000 m), selected on the basis of Lidar measurements (Mijić et al., 2013).

PBL height was obtained combining results of Lidar measurement case studies (Banks et al., 2014) and calculations using MeteoInfo (Wang, 2012) software and GDAS1 data. The missing values were interpolated using linear fit. Only full-length trajectories below 120% of the PBL height at the receptor site were used for the analysis. Raman lidar is used to detect the strongest gradient of the aerosol vertical distribution, which is attributed to the boundary layer height. Bi-axial lidar system emitted the laser radiation at 1064, 532, and 355 nm which was collected at elastic (355 nm) and Raman (387 nm) channels. The dynamic range extended from 500 m up to 15 km due to combined analog and photon counting detection mode, total laser pulse energy of 300 mJ and a 250 mm Cassegrain telescope diameter. Averaging time of the Lidar profiles was in the order of 1 min.

Results and Discussion

As shown in Table 1, highly reactive BTEX species represent a significant VOC fraction, with the highest mixing ratio of 3.35 ppb observed for toluene. Thereby, ethylbenzene, mp-xylene, and o-xylene are well correlated with gaseous oxides, thus indicating the association with combustion sources, whereas lower r values ranging from 0.35 to 0.86 for benzene and toluene are most likely the result of considerable non-traffic emissions, such as industrial solvent usage or gasoline evaporations. The notable exception among gaseous oxides is SO_2, which had little or no correlation with BTEX, which implies that its major portion originates from coal burning for heating purposes. Apart from BTEX, the average concentrations of C9 aromatics, compounds detected at m/z 83 (methyl furan, hexenal and hexanal fragments) and 85 (ethyl vinyl ketone, monoterpene oxidation products), ethanol and oxygenated species comprising acetone, acetaldehyde, acetic acid and its fragments at m/z 43 are observed to be moderately to considerably higher than expected. As previously reported by Valach et al. (2014), the elevated VOC concentrations in the winter season are to some extent caused by the shallow daytime PBL with limited dilution and mixing. Furthermore, the contribution of regional and long-range transport is shown to play an important role for the observed levels of some species (m/z 83, 85 and 87), as considered below. As regards the oxygenated species, the elevated mixing ratios are at least partly influenced by specific conditions within the canyon avenue, as well as by the low rate of photochemical reactions and longer atmospheric lifetimes.

Following the standard procedure in receptor modeling 29 species were chosen as input variables producing physically interpretable solution. The attempts to include all species failed to provide feasible results, though some of them were abundant or statistically interrelated. For instance, methanol, methanol/water cluster and one of the most important peroxides - methyl hydroperoxide, being detected at m/z 33, 51 and 49, respectively, were not included in the model solutions.

Both receptor models resolved six profiles with estimated contributions in the range from 6.3 to 27.6%. In Unmix solution, profiles attributed to vehicle exhaust (VE_U), solid fuel burning (SFB_U), and local laboratory (LL_U) were assigned to single sources, whereas the profiles associated with aged plumes (AP), gasoline evaporations (GE) and petrochemical industry (PI) were considered complex, including certain contributions of species originating from other sources, such as industrial solvents (IS), oil refinery (OR) and biogenic emissions (BE), respectively. In comparison with this, PMF-identified profiles were relatively similar, with one notable exception: local laboratory was not identified as a separate source, and instead of it additional profile related to vehicle exhaust is resolved. The results of statistical tests, model evaluation and Taylor diagrams are shown in Table 2, i.e., Figure 2. Except for certain discrepancies, Unmix-resolved profiles and related PMF counterparts are in good agreement which is confirmed by statistical correlations in the range from 0.57 to 0.89, with highest values observed for the profiles associated with gasoline evaporations and emissions from oil refinery, GE/OR_U and GE/OR_P.

Table 1. Statistical parameters for VOC-related masses [ppb], NO_x, NO_2, NO, SO_2 [μg m^{-3}] and CO [mg m^{-3}] (left), and source profiles resolved by PMF and Unmix [%] (right)

Species	Mean	Median	Min	Max	St. Dev.	LL	SFB		GE/OR		VE			PI/BE		AP	
						Unmix	PMF	Unmix	PMF	Unmix	PMF_1	PMF_2	Unmix	PMF	Unmix	PMF	Unmix
NO_x	149.7	121.0	11.2	912.4	116.2	3.6	14.7	32.8	0.3	0	0	52.7	46.6	0	0	32.3	17.0
NO_2	62.9	55.0	8.9	239.1	38.2	4.8	23.5	36.1	0.3	0	0	12.0	25.7	0	0	64.2	33.4
NO	86.8	58.7	1.2	673.3	88.3	2.9	8.8	30.8	0	0	0.3	79.5	59.6	0	0	11.3	6.7
CO	0.7	0.6	0.2	3.4	0.4	2.6	16.6	23.5	6.3	8.6	13.5	21.6	53.5	28.7	7.7	13.3	4.1
SO_2	21.5	17.4	3.9	236.5	16.7	2.9	68.9	69.5	0	19.3	0.9	7.0	8.3	6.2	0	17.1	0
Ethylbenzene	2.3	1.8	<DL	36.9	2.4	2.3	6.3	1.3	4.6	0	80.3	0	82.9	1.3	0	7.5	13.5
mp-Xylene	9.0	6.9	<DL	124.6	9.5	1.9	0	2.4	0	0	54.8	12.3	89.0	18.5	0	14.4	6.7
o-Xylene	1.9	1.5	<DL	18.5	1.7	2.0	11.6	9.2	3.3	0	72.3	12.9	78.0	0	0	0	10.8
m/z 41 (propylene)	1.9	1.7	0.3	19.9	1.4	3.4	10.2	2.8	24.7	17.3	7.1	8.1	33.0	29.5	21.7	20.4	21.9
m/z 43	4.9	4.4	1.3	28.4	2.5	2.8	8.0	3.5	31.4	18.2	4.6	0	11.7	26.7	26.3	29.3	37.5
m/z 45 (acetaldehyde)	5.0	4.3	1.3	38.9	3.5	23.4	19.1	12.8	14.6	7.7	4.1	0.1	18.3	30.7	14.4	31.4	23.4
m/z 47 (ethanol)	7.8	3.0	0.1	177.3	15.6	86.0	24.0	0	26.8	7.5	0	0	0.2	22.2	2.6	27.0	3.7
m/z 57 (MTBE)	1.9	1.6	0.3	28.8	1.9	3.5	6.7	0	24.5	17.6	8.8	7.3	31.8	38.0	29.7	14.7	17.4
m/z 59 (acetone)	7.2	5.7	2.4	30.8	5.4	5.8	0	2.5	76.1	44.1	3.6	5.4	0	0	1.3	14.9	46.3
m/z 61 (acetic acid)	4.9	4.5	1.4	26.0	2.4	2.9	9.6	6.2	31.7	25.9	3.0	0	13.8	34.3	27.2	21.4	24.0
m/z 71	0.6	0.6	0.1	5.3	0.4	2.6	15.9	9.9	21.7	25.5	5.8	5.9	29.8	40.4	29.6	10.3	2.6
m/z 73	0.9	0.8	0.2	7.9	0.5	2.0	16.9	12.6	17.9	15.9	7.7	0	26.0	36.3	28.4	21.2	15.0
m/z 75	1.3	1.0	0.3	45.3	2.1	2.4	14.7	6.9	22.8	21.5	5.0	0	20.9	42.0	35.8	15.6	12.5
m/z 79 (benzene)	1.4	1.1	0.1	7.0	0.9	1.1	18.6	15.1	5.9	13.6	14.5	11.9	44.4	49.0	25.9	0	0
m/z 81	0.8	0.7	0.1	9.9	0.8	5.5	15.4	0	31.2	21.5	1.4	4.5	12.1	43.4	57.9	4.1	3.1
m/z 83	14.4	10.2	0.4	92.4	12.0	0.8	0	0	80.8	69.5	0.0	5.1	0	9.0	9.0	5.1	20.7
m/z 85	9.4	6.7	0.2	65.1	7.9	0.7	0	0	80.8	69.1	0.1	5.3	0	8.8	9.4	5.0	20.7
m/z 87	2.5	2.2	0.5	12.4	1.5	1.5	5.8	3.1	58.9	59.4	2.3	2.8	5.3	19.2	12.8	10.9	17.9

Table 1. (Continued)

Species	Mean	Median	Min	Max	St. Dev.	LL	SFB		GE/OR		VE			PI/BE		AP	
m/z 93 (toluene)	3.4	2.6	0.2	29.7	2.6	0.9	12.5	1.1	12.0	11.9	10.6	13.7	49.5	41.8	29.0	9.4	7.6
m/z 99	0.5	0.4	0.1	2.3	0.3	3.2	25.7	18.8	30.1	39.5	0	0	8.2	43.1	30.3	1.2	0
m/z 101	1.3	1.1	0.4	5.3	0.7	4.3	18.5	18.8	40.9	52.4	0	0	0.7	31.5	23.8	9.2	0
m/z 105 (styrene)	0.5	0.4	0.1	16.4	0.8	1.5	12.8	8.7	21.8	24.7	4.6	6.1	29.2	44.6	31.6	10.0	4.1
m/z 121 (C9 aromatics)	4.3	3.1	0.3	25.0	3.9	0.1	19.3	4.3	29.4	65.8	1.3	8.2	20.0	41.8	9.9	0	0
m/z 137 (monoterpenes)	0.9	0.7	0.0	8.5	0.9	6.5	16.2	0	39.2	27.0	0	0.1	2.6	44.5	63.9	0	0
DL - detection limit			Average contribution:			**6.3**	**14.5**	**11.5**	**25.4**	**23.6**	**10.6**	**9.7**	**27.6**	**25.2**	**18.2**	**14.5**	**12.8**

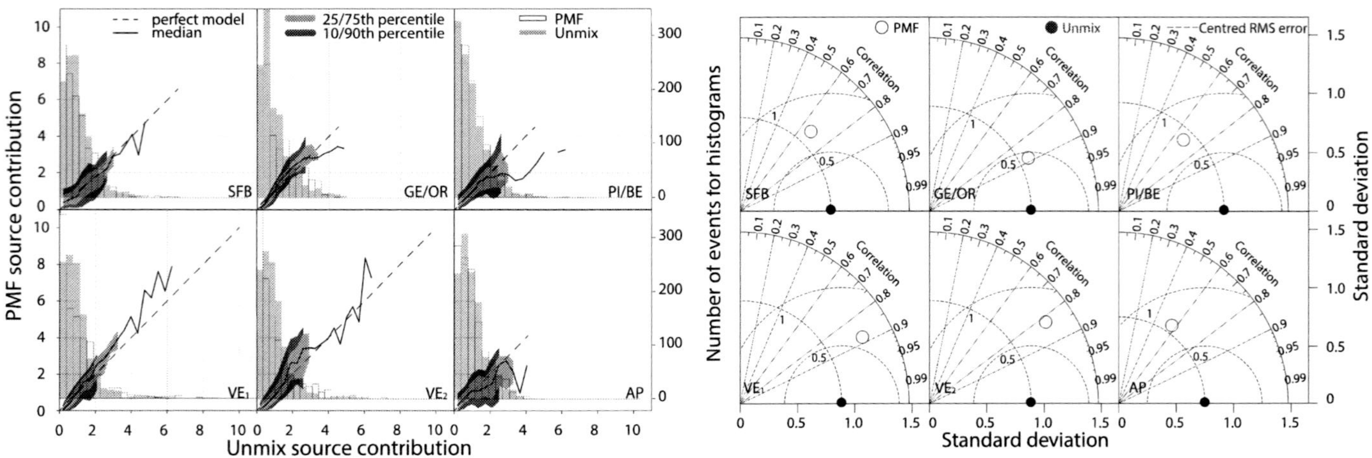

Figure 2. Model evaluation (left) and Taylor diagrams (right).

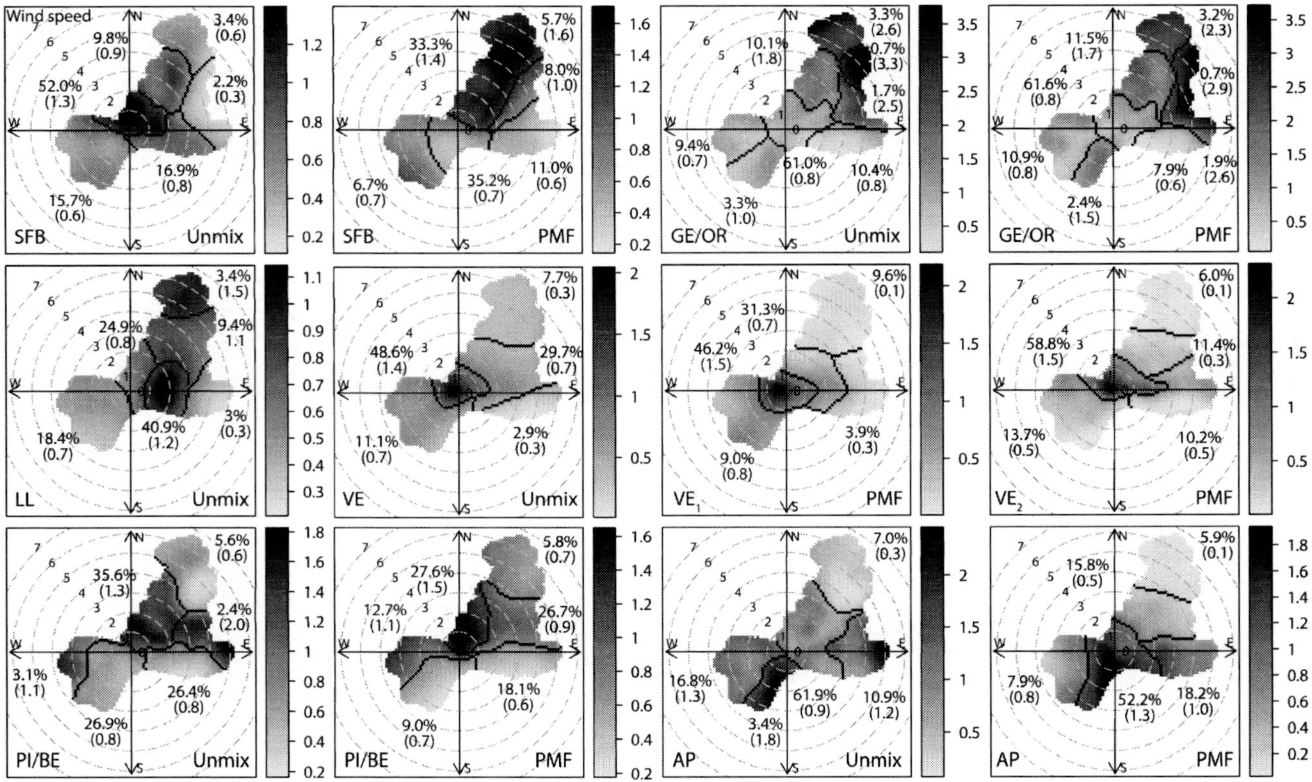

Figure 3. The relationship between normalized source contributions and wind characteristics, together with bivariate cluster plot (percentage and average contribution).

Table 2. Evaluation statistics for assessing PMF/Unmix model performance

Source	FAC2	NMB	NMGE	RMSE	COE	IOA
SFB	0.6005	-0.0054	0.5051	0.6832	0.1384	0.5692
GE/OR	0.7648	0.0030	0.3086	0.4413	0.5304	0.7652
VE_1	0.6689	0.0161	0.4012	0.5908	0.3344	0.6672
VE_2	0.5766	0.0119	0.4915	0.7045	0.1846	0.5923
PI/BE	0.6809	0.0002	0.4661	0.6917	0.3043	0.6522
AP	0.6382	0.0004	0.5311	0.7228	0.0828	0.5414

It should be mentioned that herein presented six-profile solution appeared to offer the most feasible and interpretable results in both models. The attempts to reduce the number of profiles to four or five resulted in merging of Unmix profiles of different sources located in the industrial area of Pančevo. Furthermore, the major portions of BTEX in PMF solution collapsed into one profile, resulting in an unrealistic estimation of source contributions. In the extended model solutions, vehicle exhaust profiles were excessively divided and thus, the main portions of traffic-related species and gaseous oxides, suitable indicators for anthropogenic combustion processes, were apportioned into separate profiles. In addition, the correlations between Unmix and PMF profiles were highest for six-profile solutions.

As can be observed, no profile attributed exclusively to natural emissions was resolved, even though Botanical Garden with a large biodiversity and ornamental plants is located approximately 150 m NE from the measurement site. Moreover, the major biogenic tracer, isoprene at m/z 69, was not apportioned to any of the resolved profiles. The average isoprene concentration of 0.79 ppb was higher than expected, probably due to the emissions from the vegetation area, even though significant man-made contribution to isoprene levels has also been reported in urban traffic sites (Schneidemesser et al., 2011). Generally speaking, the emissions from biosphere are negligible in urban environments, particularly during the winter season. Nevertheless, their impact on urban air quality has been recently emphasized due to the urban-heat-island effect and high reactivity of biogenic tracers in complex oxidative atmospheric conditions with the abundance of precursors. For instance, in densely vegetated regions, relatively fast reaction rates of biogenic species can result in ozone formation substantially greater than in the cities with large transportation sources (Baker et al., 2008).

Local Laboratory

As previously mentioned, laboratory emissions were identified as a separate source only in Unmix solution, with the lowest estimated contribution of 6.3%. Bivariate cluster plot reveals that the dominant portion (64%) of the profile contribution is of strictly local origin (Figure 3). The chemical composition of LL_U is distinguished by the major portion of common solvent and antiseptic, ethanol (86.0%), moderate portion of acetaldehyde (23.4%) and negligible contributions of other species. In addition to solvent usage in the local laboratory, the results of both PMF and Unmix show that the higher average concentration of acetaldehyde (5.03 ppb) is also associated with the emissions from vehicle exhaust, solid fuel burning and suburban industrial area. Due to the fact that it is emitted from both local and remote sources, no significant variations in acetaldehyde levels were observed during the episode of extreme wind. At the same time, concentrations of locally generated ethanol appeared to be lower than usual with rather small, sharp and narrow peaks observed as a result of the enhanced dispersion. As regards profile dynamics, diurnal pattern remains consistent during the weekdays, with a gradual increase coinciding with the working activity from early morning till broad evening maximum (7:00 AM – 6:00 PM) (Figure 4). The similar pattern is observed on Saturdays with markedly less intense peaks especially in the afternoon, whereas the accumulated pollutants are diluted to minimum overnight and on Sundays. It is also worth noting that acetonitrile and ethanol have almost identical daily and weekly profiles, together with disinfectant formaldehyde, which is intentionally excluded from the dataset due to back reactions with water vapor, and hence, low detection sensibility. The noticeable decrease from January to March is probably associated with few short-term episodes of higher evaporative emissions registered at the beginning of the measurement period. Besides these sporadic incidents, the profile contribution was relatively uniform during weekdays.

Solid Fuel Burning

The profiles interpreted as solid fuel burning are distinguished by the dominant portion of SO_2 (68.9-69.5%) and significant contributions of the associated gaseous oxides, NO_x, NO, NO_2 and CO, which have more than one source of origin. Unmix and PMF profiles are statistically interrelated (0.68) and have similar compositions. However, the share of VOC species and combustion gases other than SO_2 for the two model solutions are slightly different, thus resulting in different contribution estimates (11.5 and 14.5%, respectively).

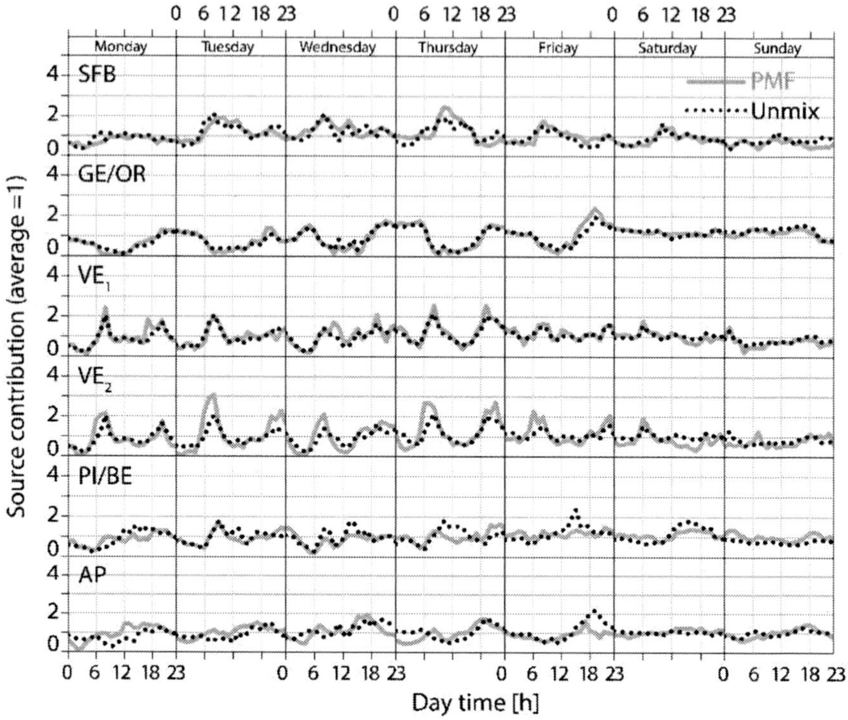

Figure 4. Diurnal variations of PMF- and Unmix-resolved normalized source contributions.

Unlike in Western Europe, the energy sector in the South-East Europe is dominantly based on coal-fired power plants that remain the largest sources of SO_2 emissions, followed by different industrial processes. Most of the thermal power plants in Serbia are lignite-fired, without the proper air pollution control equipment. Due to its high moisture and low energy content, lignite is considered the lowest rank of coal, producing highest pollutant loadings during combustion. Unfortunately, flue-gas desulfurization system, as the most frequently used technology for SO_2 reduction, will probably not be installed in the near future since it requires considerable financial investment and changes in the currently operating units. Additionally, roughly 50% of the urban households in Serbia are connected to the district heating system, or have central heating via furnaces in their homes or buildings. Besides that, a significant percentage of households, even in Belgrade, rely on the combustion of solid fuels such as coal and crude oil, which severely impairs air quality. Having that in mind, the estimated contributions of solid fuel burning in the winter season would be expected to exceed 15%. However, neither thermal plants nor local fireboxes were operating at their full capacity since the average temperature of 8.5°C in winter 2014 was considerably higher than usual. Following the temperature rise in March the contribution of both profiles rapidly decreased, whereas certain differences in temporal dynamics of gaseous oxides were evident. The end of the heating season in March was marked by a significant decrease in SO_2 mixing ratios. During the same period, the change in CO level is negligible, whereas the concentrations of nitrogen oxides with major shares apportioned between vehicle exhaust and chemical industry are observed to exhibit a continuous increase.

The contributions of several other species that are considered to be of anthropogenic (benzene, toluene, acetaldehyde, methyl ethyl ketone (MEK), methyl vinyl ketone (MVK), styrene and C9 aromatics) and biogenic origin (monoterpenes and species detected at m/z 81, 99, 101) were observed to be low to moderate, depending on the model solution. For both profiles, calculated toluene-to-benzene ratios (T/B) were of similar order to previously published estimates, derived for biofuel, wood and charcoal burning (Johansson et al., 2004). As regards isoprenoid species apportioned to PMF profile (SFB_P), their origin can be associated with the vicinity of the large vegetation area, but also with wood consumption for residential heating.

In the presented solutions, acetonitrile was not apportioned to any of the resolved profiles, although the concentrations of 3.37 ppb registered at the measurement site considerably exceed the levels in the free troposphere (0.1–0.2 ppb) (Zheng et al., 2013). Acetonitrile is often used as a tracer for biomass-burning, even though traffic-related emissions in urban areas may also explain a substantial portion of its contribution. However, diurnal profile of acetonitrile is in this case characterized by steady increase in the afternoon and evening hours, which reflects fuel burning rather than vehicle exhaust emission pattern. Therefore, we would assume that the major sources of acetonitrile can be assigned to household biofuel usage and/or fossil-fuelled thermal power facilities, although it remained unrecognized in the model solutions.

The CBPF plot above the 75th percentiles clearly reveals the dominance of the local emissions in both Unmix and PMF profiles, whereas in the percentile ranges from 0 to 50, the impact of diluted plumes from more distant sources is observed (Figure 5). In compliance with this, bivariate cluster analysis for SFB_U showed that approximately 52% of the profile contribution could be associated with the emissions from the surrounding residential sector.

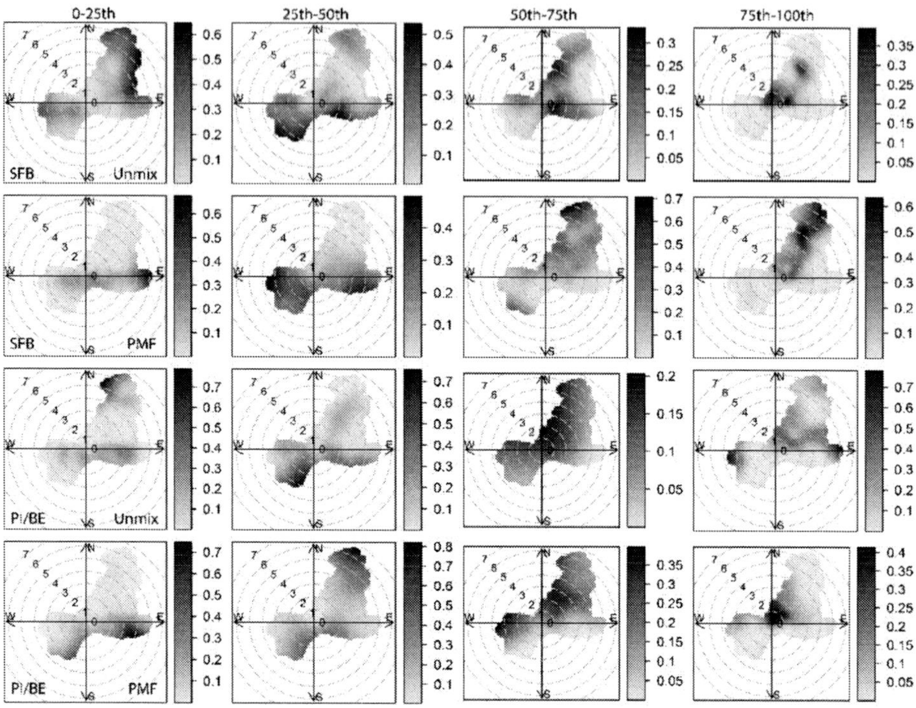

Figure 5. CBPF plot in four ranges of percentiles for PI/BE (above) and SFB sources (below).

As reported by Barletta et al. (2005), processes such as residential coal burning, where the fuel is consumed at lower temperature under smouldering conditions, are less efficient. Therefore, they appear to be emission sources of higher importance at least on a local level, compared to the combustion in thermal plants and industrial facilities that dominates on a global scale. In the case of SFB_P, the bivariate cluster plot showed considerable contribution (33%) from the N/NE sector, whereas the impact of local sources is estimated to be lower than for its Unmix counterpart (35.2%). The observed contribution from N/NE was detected on two occasions: during the episode of strong NE wind due to the transport of C9 aromatics, as well as during the prevalence of the N/NE wind of moderate speed due to the enhanced transport of biogenic species (monoterpenes and their oxidation products at m/z 81) from Botanical

Garden. According to Holzinger et al. (2005), monoterpenes rapidly decay within few minutes, thus suggesting that their maximum concentrations cannot be significantly spatially separated from the primary emission source.

Daily variations of contributions are similar for both SFB profiles and characterized by a pronounced morning maximum (6:00 - 11:00 AM) and lower evening peak (7:00 - 10:00 PM) (Figure 4). Both SFB contributions exhibited noticeable decrease during a strong wind episode, which confirms that the majority of the contributing species are of local origin. It should be emphasized that the shifts between high and low pollutant levels are not correlated with the changes in wind direction and speed only in the case of pollutants with numerous sources and long atmospheric lifetimes, since their mixing ratios are homogenous over large areas.

Gasoline Evaporations/Oil Refinery

In their review article dealing with VOC source apportionment in 20 urban areas mostly in the United States, Watson et al. (2001) concluded that gasoline vehicle exhaust, liquid gasoline and gasoline evaporations contribute up to 50% or more to the ambient concentrations. In this chapter, evaporative losses through the tank vent, leakage and permeation during fuel refining, transportation and storage are also shown to present a significant source of VOC pollution, besides the emissions related to fuel combustion processes. Due to their complexity and abundance of different pollutants, the presented profiles are attributed to gasoline evaporations and emissions from oil refinery. Furthermore, the presence of oxygenated species, such as acetone, and the highest observed contribution of transported pollution clearly indicate that the profiles also reflect source emissions from remote areas.

The estimated contributions are comparable and amount to 23.6% and 25.4%. for GE/OR_U and GE/OR_P, respectively. Profile compositions are similar, with acetone and C9 aromatics being discernible exceptions. The first one contributing with 76.1% to GE/OR_P is, in Unmix solution, equally apportioned between GE/OE_U and aged plumes. Conversely, considerable share of C9 aromatics (65.8%) apportioned to GE/OR_U, is partitioned and assigned to petrochemical industry and solid fuel burning in PMF solution. Additionally, modest discrepancies between the resolved profiles arise from the estimated shares of several other species: in GE/OR_P higher contributions of ethanol, monoterpenes and species at m/z 43 (acetic acid fragments), 83 (methyl furan, hexenal and hexanal fragments) and 85 (ethyl vinyl ketone, monoterpene oxidation products) are observed, whereas GE/OR$_U$ contains higher portions of SO_2 and species detected at m/z 99 (hexenal, monoterpene oxidation products) and 101 (hexanal). Despite the listed differences, the statistical correlation between the profiles of 0.89 is the highest calculated.

According to the results of TSA, the profiles are marked by considerable contributions of regional transport, being 27.5% and 30.5% for GE/OR$_U$ and GE/OR$_P$, respectively. The chemical compositions of the profiles are complex, comprising two groups of species which significantly differ in terms of their contributions, temporal dynamics and origin.

The first group containing acetone, C9 aromatics, and species detected at m/z 83, 85, 87, 99, and 101, is distinguished by their major portions (up to 81%) and high correlations with the profiles (0.43-0.91). According to the overview of VOC sources in the literature, the signals registered at m/z 83, 85 and 87 are often attributed to biogenic species, e.g., hexanal, methyl furan, 2-methyl-3-buten-2-ol and unidentified monoterpene oxidation products, while in our case, two details speak in favour of their man-made origin. Firstly, the average concentrations ranging from 2.53 to 14.40 ppb significantly exceed the expected concentrations of biogenic VOC in winter season. Furthermore, they exhibit higher correlations with petroleum volatiles, C9 aromatics (0.58-0.61) than with monoterpenes (0.36-0.39), and thus, it is more likely that the observed signals result from industrial processes, e.g., cyclopentane (m/z 83) and thiophene (m/z 85) are commonly found in petroleum evaporations, whereas 2-pentanone and pentanal (both at m/z 87) are used in chemical industry. Similarly, the signals at m/z 99 and 101 are often attributed to hexenal and hexanal, although their correlations with C9 aromatics (0.53-0.57) and biogenic tracers (0.48-0.58) are comparable.

Acetone has some contributions from anthropogenic and biogenic sources; however the secondary formation from the atmospheric oxidation of hydrocarbons is reported to be the single largest source with 51% of the global acetone budget (Singh et al., 1994). Precursor species include propane, found in liquid petroleum gas, as well as isobutane and isobutene, as the major feedstock in the production of synthetic rubber monomers. According to the same author, the remaining portion of the global source can be attributed to biomass burning. Nevertheless, in this chapter, acetone is apportioned between GE/OR and profile identified as aged plumes, with the dominant share assigned to GE/OR in PMF solution. The absence of correlations with gaseous oxides and monoterpenes confirms that acetone levels are neither dependent on combustion processes, nor on biogenic emissions, but rather on atmospheric oxidation of transported or locally emitted species.

The second group comprises species with moderate contributions such as propylene, MTBE, styrene, acetic acid, and compounds detected at m/z 43 (acetic acid fragments), 71 (MVK, methacrolein (MACR), isoprene oxidation products), 73 (MEK, butanal), and 75 (hydroxyacetone, propionic acid), possibly originating from gasoline, liquid petroleum evaporations and chemical industry. Apart from the listed anthropogenic species, monoterpenes and their oxidation products at m/z 81 contribute with certain share as well.

Since the results of PSCF and CWT analyses point to the same conclusions for both profiles only CWT plot for GE/OR$_U$ is presented here, to avoid redundancy (Figure 6). As can be seen, the transported species originate from wider areas, with the strongest emission sources located in the NE, followed by those in E and W sectors. The results of LIDAR measurements show that diurnal variations in shallow winter PBL height ranged from 500 to 1500 m.

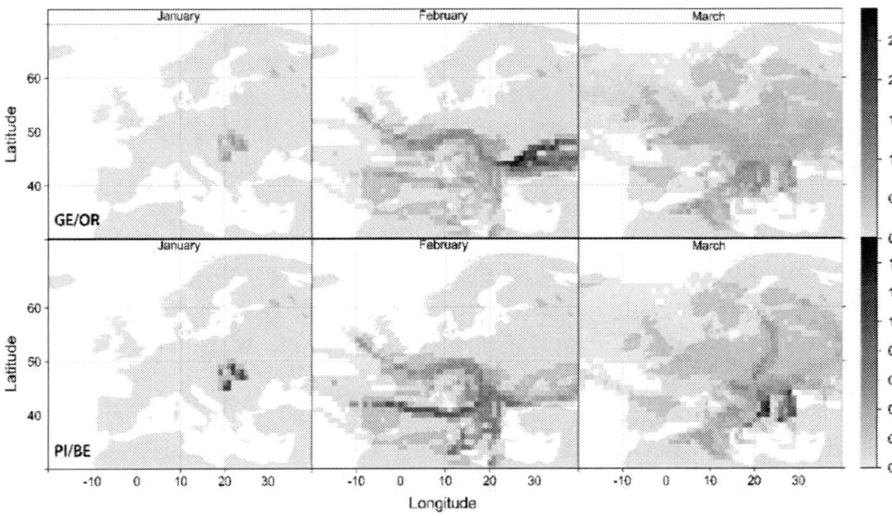

Figure 6. CWT-derived maps for Unmix resolved GE/OR and PI/BE normalized source contributions.

Lower PBL height of nearly 500 m in January enabled regional transport from Hungary, Romania and Slovakia, whereas with the increase of daily PBL height to 900 m in February, the area of incoming air masses was expanded towards NW and NE/E area, mostly Romania. Finally, further increase of PBL height to around 1500 m and change of meteorological conditions in March resulted in somewhat lowered impact of long-range pollutant transport from

S/SE. On its pathway to the measurement site in January and February, the air masses were constantly encompassing oil and gas refinery in the industrial area of Pančevo, located around 8 km NE of the Belgrade downtown edge. With the annual production of 1.6 million tons of crude oil and 3.1 million tons of the related petroleum products including gasoline and diesel fuel, fuel oil, kerosene, liquefied petroleum gas, the refinery in Pančevo is one of the largest oil producers in the South-East Europe and an important regional source of air pollution.

Contributions of both GE/OR_U and GE/OR_P display similar diurnal patterns: the midday minimum on working days is followed by a steady increase in the evening hours to the level which remains fairly constant over the night (Figure 2). The observed noontime decline can be associated with air temperature increase from 6:00 to 12:00 AM and the following photochemical removal, although atmospheric reactions in winter season are much slower due to lower solar intensity and shorter day length (Kumar Padhy and Varshney, 2000). The evaporations accumulated during the afternoon are trapped with the nocturnal inversion (Allan et al., 2010) and diluted with the increase in PBL height during the following morning. As regards contributions on Saturday and Sunday, relatively high levels are typical for evaporative emissions rather than for emissions resulting from anthropogenic activities. However, the absence of midday decline during the weekend is evident and can be associated with long-range transport which occurred once with the strong wind episode at the end of the week (from January 30th, Thursday to February 2nd, Sunday), and affected the entire period. Thereby, the rapid rise in the concentrations of all major species was observed, with the exception of that registered at m/z 99 (hexenal, monoterpene oxidation products). Another possible reason for stable weekend GE/OR levels is that a decrease in industrial- and traffic-related emissions on Saturday and Sunday resulted in lower atmospheric oxidizing ability, which further affected the rate of photochemical reactions. However, it should also be mentioned that emissions from remote sources are not expected to exhibit clear patterns, particularly on a daily basis, due to pollutant dispersion with spatial and temporal distance.

Moreover, the intense four-day transport episode at the beginning of the measurement period had probably an impact on monthly dynamics as well, thus making the contributions appear like decreasing from January to March. Besides that, the PBL height increase to 1500 m in March affected the stability of lower troposphere, which together with the change in winter direction from NE/E to SW/W led to the enhanced dispersion of pollutants at the measurement site.

Vehicle Exhaust

Nowadays, traffic-related emissions are the leading VOC source, and a more serious environmental problem in the developing countries than in Western Europe and the USA (Han and Naeher, 2006). According to the same study, traffic volume and the percentage of traffic jam accounted for even 0.76 to 0.80 of the variability in BTEX concentrations. As regards Serbia, the Statistical Office reported that about half a million of cars and 120,000 other types of motor vehicles were registered in 2011 in the urban area of Belgrade, thus making the vehicle exhaust emissions one of the biggest environmental threats.

The slow and congested traffic in central city areas severely impairs air quality, particularly during the cold season, when VOC emissions from the start-up phase can amount to thousands of driven kilometres (Nordin et al., 2013). In their study dealing with VOC pollution in Delhi, Kumar Padhy and Varshney (2000) confirmed that vehicle exhaust contains several times higher concentrations of unburnt hydrocarbons during idling and deceleration (500 - 1000 and 3000 - 12,000 ppm, respectively) than during acceleration and cruising (50 - 800 and 200 - 800 ppm, respectively). Thereby, newer engines are even more likely to contribute to high VOC levels due to incomplete combustion caused by fuel condensate on the inner surface of the cylinder, as well as because the temperature for oxidation catalyst operation is well below its optimum range in winter.

Unmix-derived profile interpreted as vehicle exhaust (VE_U) with estimated average contribution of 27.6%, was in the PMF solution divided into two separate profiles, VE_{P1} and VE_{P2}, with total contribution being considerably lower (20.3%) (Stojić et al., 2015b). The profiles are enriched in o-xylene, mp-xylene, ethylbenzene (with contributions up to 89.0%), and gaseous oxides, except for SO_2 which was previously shown to originate mainly from solid fuel burning. Additionally, the considerable shares of benzene, toluene and other traffic-related species are apportioned to VE profiles with contributions being almost twice higher in VE_U. As regards this, C9 aromatics and methyl tertiary butyl ether (MTBE) with moderate shares apportioned to VE_U (20% and 31.8%, respectively) are according to the PMF solution, mostly registered as a result of gasoline evaporations and emissions from petrochemical industry. Similarly, the contributions of the species detected at m/z 71 and m/z 73 are approximately three times higher in Unmix profile, whereas PMF suggests that their source of origin is mainly petrochemical industry. In the literature, the signals at m/z 71 and m/z 73 are commonly associated with

MEK, MACR, MVK and other unidentified products of monoterpene and isoprene oxidation, which are emitted from both biogenic and anthropogenic sources. In compliance with this, it should be noted that correlations between the species at m/z 71 and m/z 73 and anthropogenic pollutants, such as propylene, MTBE, benzene, toluene, and styrene, are in the range from 0.64 to 0.89, which is comparable with the correlations they exhibit with biogenic tracers, monoterpenes and isoprene (from 0.47 to 0.91). Therefore, we would assume that the signals at m/z 71 and m/z 73 are detected as a result of both vehicle exhaust and emissions from vegetation, and their origin can be recognized as biogenic in the profiles related to petrochemical industry, which contain higher amounts of monoterpenes. Due to the observed discrepancies in contribution levels of several species, model-estimated contribution of petrochemical industry also differs, and is about 6.3% higher for PMF-resolved profile than for its Unmix counterpart. Despite the observed differences, VE_U, VE_{P1} and VE_{P2} are highly correlated with $r > 0.8$.

The calculated T/B ratios ranging from 2.16 and 2.49 are in good agreement with the previous findings of Leucher and Rappenglück (2010). Similarly, in their study on traffic-related emissions in the cold weather season, Lough et al., (2005) confirmed that T/B ratios were in the range from 2 to 5.

The previous findings suggest that the number of gasoline vehicles significantly exceeds that of diesel, at least in the urban zones, with dominantly large city buses running on diesel fuel (Watson et al., 2001). Unlike gasoline, diesel emissions are associated with 15 times less CO, seven times more NO_x and 10-14 times less BTEX (Thornhill et al., 2010). The similar shares of CO, NO_x in the range from 35.1% to 53.5%, and 46.6% to 52.7%, respectively, together with considerable portions of benzene, toluene, and extremely high portions of o-xylene, mp-xylene and ethylbenzene confirm that the majority of the observed pollution is related to emissions from gasoline engines.

The bivariate cluster analysis reveals that vehicle emissions, mainly originating from the local hotspot area (55.0% and 64.4%), are slowly diluted across the canyon avenue with low ventilation (Figure 3). A small contribution of 11% from the W sector coinciding with the wind of low speed can be assigned to the neighboring congested crossroad. Both profiles exhibit negative correlations with the wind speed in the range from -0.47 to -0.42, which further confirms the source local origin.

As regards VE profile dynamics, a clear daily pattern reflecting double rush hour, with a prominent morning peak and flattened late afternoon increase

(6:00 - 09:00 AM, 6:00 - 10:00 PM) is observed (Figure 4). During afternoon (12:00 – 05:00 PM) a significant decrease in contributions occurs, possibly due to sparse traffic flow and the boundary layer expansion. The observed weekday/weekend difference was expected as a result of decreased traffic intensity in the central city area. Higher levels of traffic-related pollutants are commonly registered during the cold weather season since the removal through photochemical reactions is reduced. However, the analyzed profile contributions exhibit a gradual increase towards the end of the measurement period, probably due to the shift in wind direction, from the NE in January and February to more polluted E, SW and W sectors in March. It should also be noted that dilution effect of the southern winds blowing transversely to the street may play a minor role in the VOC concentrations.

Petrochemical Industry/Biogenic Emissions

Besides the oil refinery and nitric plant, Petrohemija, as the largest facility of its kind in the region with the annual production exceeding 600,000 t, is also located in the industrial area of Pančevo around 13 km NE/E from the measurement site. The wide range of its products comprises propylene, commonly used fuel oxygenate (1,3-butadiene), fuel additive (MTBE), plastic and rubber polymers, as well as other related petrochemicals. The results of bivariate cluster analysis confirm that between 72.7% and 81.5% of pollution assigned to the petrochemical industry was coinciding with the wind of low to moderate speed blowing from the N/E sector. Additionally, minor contributions of air masses from SW are also observed, probably as a result of two incidents in March, when extreme concentrations of several compounds were detected.

As can be seen in Figure 4, a significant decrease in both profile contributions was registered at the end of the measurement period, although they would be expected to increase due to the rise in temperature and consecutive intensified evaporation. We assume that this might be the result of the shift in wind direction. Namely, the monthly averaged wind rose indicates (diagram not shown) that the eastward wind (40%) occurred during the entire period, whereas the second prevailing wind direction changed from N/NE in January and February to SW/W in March, which further led to a decrease in contribution from the northeastern industrial area.

The estimated impact of industrial emissions on air quality in Belgrade central zone differs for Unmix (PI/BE_U) and PMF (PI/BE_P), although the profiles are well correlated, particularly in case of source contributions being less than 1. As regards the chemical composition, the profiles are enriched in propylene, MTBE, benzene, toluene, styrene, species detected at m/z 71 (MVK, MACR, isoprene oxidation products) and 73 (MEK, butanal), and a commonly used solvent, propionic acid (m/z 75), with contributions being slightly higher in PI/BE$_P$. Most of the listed species originate from a number of different sources, out of which the petrochemical and chemical industry account for a significant portion on a global level. For instance, vehicle exhaust contains only a small amount of styrene, whereas its' largest emission source is reported to be the industry of plastic, synthetic rubber and resins (U.S. EPA, 1994).

The considerable shares of acetic acid, acetaldehyde, and acetic acid fragments detected at m/z 43 could be associated with the air masses regularly transiting over Methanol and Acetic Acid Complex, located 100 km NE from the measurement site. With the majority of species originating from evaporative emissions, no contributions of the gaseous oxides are observed, except for certain share of CO being apportioned to the PI/BE$_P$.

The results of TSA reveal that the average contribution of the regional and long-range transport amounted to 23 and 27%, with prominent increase from January to March. Therefore, the regional transport of air masses from Greece and long-range transport from NE had significant impact on the air quality at the measurement site in March, together with the local emissions from petrochemical industry (Figure 6).

As regards PI/BE dynamics, diurnal pattern clearly reflects anthropogenic activities on weekdays, with a steady increase during working hours (9:00 AM – 6:00 PM), broad afternoon-evening plateau and the nighttime minimum (Figure 4). On a weekly basis, the observed rise in profile contributions from Monday to Friday is followed by a gradual decrease on Saturday and Sunday.

It should be noticed that, in spite of the fact that Botanical Garden is located in the very vicinity of the measurement site, no profile is solely assigned to biogenic emissions. Such apportionment is consistent with the previously mentioned general findings regarding the minor role of biogenic emissions in urban environments. Nevertheless, several VOC species that might originate from biosphere, such as those detected at m/z 71 (MVK, MACR, isoprene oxidation products), m/z 73 (MEK, butanal), m/z 99 (hexenal, monoterpene oxidation products), and m/z 101 (hexanal) are assigned to PI/BE profiles, together with major shares of monoterpenes and their oxidation products at m/z 81, due to the fact that Botanical Garden is located on the pathway of air masses coming from the petrochemical industrial complex. It should also be mentioned that isoprene and monoterpenes, commonly used as biogenic tracers, exhibit significant correlations only with PI/BE_U (0.61 and 0.73, respectively), and PI/BE_P (0.44 and 0.43, respectively), which confirms that modest contribution from the biosphere is associated with the analyzed profiles.

Aged Plumes

As previously described, the measurement site is located in a boulevard with high daytime traffic density and multistoried buildings on both sides of the street, which considerably limits pollutant dispersion. In such stagnant conditions, the detection of species at different stages of evolution, from those newly formed to oxidation products with longer lifetimes, is normally expected. In compliance with this, both Unmix (AP_U) and PMF (AP_P) profiles attributed to aged plumes comprise moderate to high amounts of oxygenated VOC, including acetone, hydroxiacetone (m/z 75), acetaldehyde, acetic acid and its fragments (m/z 43), and MEK (m/z 73). Additionally, the negligible shares of gaseous oxides are apportioned to both profiles, with the exception of NO_2, which is the only abundant with major contributions of 33.4 and 64.2%. Unlike other oxides, NO_2 is known to be predominantly a secondary air pollutant, with only a small percentage originating from the primary emissions, and according to Zheng et al. (2013), it could be even treated as a reliable tracer for aged plumes.

The profile contributions are estimated to be relatively similar (12.8% and 14.5%), although noticeable differences are evident in the chemical compositions. Thereby, the higher shares of ethanol, acetaldehyde and gaseous oxides are apportioned to AP_P, whereas AP_U contains more acetone, acetic acid fragments (m/z 43) and species detected at m/z 83 (methyl furan, hexenal and hexanal fragments), 85 (ethyl vinyl ketone, monoterpene oxidation products) and 87 (methylbutenol, methyl isobutyl ketone). Even if the species at m/z 83, 85 and 87 would be considered biogenic, which is unlikely due to their high concentrations, they could not be attributed to fresh emissions from biosphere.

Intense solar radiation, higher temperatures and stagnant meteorological conditions in March affected the rate of photochemical reactions resulting in the increase of profile contributions (Figure 4). Diurnal patterns of both profiles were shown to follow the daytime release of primary pollutants, with the maximum attained at 06:00 PM, and subsequent decline in the late evening and overnight. Furthermore, lower levels of potential atmospheric precursors on Saturday and Sunday resulted in significant weekday/weekend difference. As regards spatial distribution, the polar plots confirmed that the majority of the species originate from the sources highly localized in the surrounding of the measurement site.

Additionally, the moderate contributions from the eastern sector could be attributed to the emissions from the chemical plant, located in the industrial area of Pančevo.

CONCLUSION

This chapter presents source apportionment results of the winter season from the canyon avenue with severe traffic congestion during rush hours. The resolved profiles were attributed to: vehicle exhaust, gasoline evaporations/oil refinery, petrochemical industry/biogenic emissions, aged plumes, solid fuel burning, and local laboratory. In spite of certain discrepancies between PMF and Unmix results, the compared profiles are statistically interrelated.

Apart from the traffic-related emissions that are widely recognized as a major environmental threat in urban areas, the significant impact of industrial emissions from suburban area and transported pollution from remote sources is also observed and analyzed.

According to the results of TSA, lower PBL height at the beginning of the measurement period enabled regional transport from Hungary, Romania and Slovakia. The area of incoming air masses was expanded towards NW and NE/E, mostly Romania, in February, while the increase of PBL height at the end of the measurement period resulted in long-range transport from S/SE. It should also be noticed that no model solution resolved a profile attributed exclusively to biogenic emissions, even though Botanical Garden is located in the vicinity of the measurement site.

In conclusion to this, the presented results might provide useful suggestions for design and implementation of environmental regulation measures aimed at reducing the community exposure. Nevertheless, further analyses are required for better understanding of the impact of local factors and transported pollution on VOC ambient concentrations and their evolution in the region.

ACKNOWLEDGMENTS

This paper was realized as part of projects No III43007 and No III41011, founded by the Ministry of Education, Science and Technological Development of the Republic of Serbia for the period 2011-15. The authors are thankful to the Institute of Public Health in Belgrade for providing the necessary facilities and data.

REFERENCES

Allan, J. D.; Williams, P. I.; Morgan, W. T.; Martin, C. L.; Flynn, M. J.; Lee, J.; Coe, H. *Atmos Chem Phys.* 2010, *10*, 647-668.

Aničić, M.; Frontasyeva, M. V.; Tomašević, M.; Popović, A. *Envir Monit Assess.* 2007, *129*, 207-219.

Aničić, M.; Spasić, T.; Tomašević, M.; Rajšić, S.; Tasić, M. Ecol Indic. 2011, 11(3), 824-830.

Aničić, M.; Tasić, M.; Frontasyeva, M. V.; Tomašević, M.; Rajšić, S.; Mijić, Z.; Popović, A. *Environ Pollut.* 2009, *157*, 673-679.

Baker, A. K.; Beyersdorf, A. J.; Doezema, L. A.; Katzenstein, A.; Meinardi, S.; Simpson, I. J.; Sherwood Rowland, F. *Atmos Environ.* 2008, *42*, 170-182.

Banks, R. F.; Tiana-Alsina, J.; Baldasano, J. M.; Rocadenbosch, F. *Proc. SPIE 9242, Remote Sensing of Clouds and the Atmosphere XIX; and Optics in Atmospheric Propagation and Adaptive Systems XVII*; International Society for Optics and Photonics, 2014, pp. 92420F-92420F

Barletta, B.; Meinardi, S.; Sherwood Rowland, F.; Chan, C. Y.; Wang, X.; Zou, S.; Blake, D. R. *Atmos Environ.* 2005, *39*, 5979-5990.

Barley, M. H.; Topping, D.; Lowe, D.; Utembe, S.; McFiggans, G. *Atmos Chem Phys.* 2011, *11*, 13145-13159.

Belis, C. A.; Karagulian, F.; Larsen, B. R.; Hopke, P. K. *Atmos Environ.* 2013, 69, 94-108.

Brankov, E.; Henry, R. F.; Civerolo, K. L.; Hao, W.; Rao, S. T.; Misra, P. K.; Reid, N. *Environ Pollut.* 2003, *123*, 403-411.

Brown, S. G.; Frankel, A.; Hafner, H. R. *Atmos Environ.* 2007, *41*(2), 227-237

Carslaw D.C.; Ropkins K. *Environ Modell Softw.* 2012, *27,* 52-61.

Carslaw, D.C.; Beevers, S.D. *Environ Modell Softw.* 2013, *40*, 325-329.

Draxler, R. R. Rolph, G. D. (2014). HYSPLIT (HYbrid Single-Particle Lagrangian Integrated Trajectory) model access via NOAA ARL READY website (http://ready.arl.noaa.gov/HYSPLIT.php). NOAA Air Resources Laboratory, Silver Spring.

Elbir, T.; Cetin, B.; Cetin, E.; Bayram, A.; Odabasi, M. *Environ Monit Assess.* 2007, *133*, 149-160.

Galbally, I. E.; Lawson, S. J.; Hibberd, M. F.; Bentley, S. T.; Cheng, M.;, Weeks, I. A. Dunne, E. (2008). A study of VOCs during winter 2006 at Wagerup, Western Australia. A Report to Alcoa World Alumina Australia.

Han, X.; Naeher, L. P. *Environ Int.* 2006, *32*, 106-120.

Hare, J. D. *Annu rev entomol.* 2011, *56*, 161-180.

Henry, R. C.; Christensen, E. R. *Environ Sci Technol.* 2010, *44*, 2474-2481.

Henry, R.C. (2003). Multivariate receptor modeling by N-dimensional edge detection. *Chemometr Intell Lab* 65, 179-189.

Holzinger, R.; Lee, A.; Paw, K. T.; Goldstein, U. A. H. *Atmos Chem Phys.* 2005, *5*, 67-75.

Hsu Y. K.; Holsen T. M.; Hopke P. K. *Atmos Environ.* 2003, *37*, 545-562.

Huang, Y.; Ho, S. S. H.; Ho, K. F.; Lee, S. C., Yu, J. Z.; Louie, P. K. *J Hazard Mater.* 2011, *186*(1), 344-351.

Huber, M.; Knutti, R. *Nat Geosci.* 2012, *5*, 31-36.

Johansson, L. S.; Leckner, B.; Gustavsson, L.; Cooper, D.; Tullin, C.; Potter, A. *Atmos Environ.* 2004, 38, 4183-4195.

Jung, J. H.; Choi, B. W.; Kim, M. H.; Baek, S. O.; Lee, G. W.; Shon, B. H. *Environ Health Toxicol.* 2012, 27.

Kansal, A. *J Hazard Mater.* 2009, *166*(1), 17-26.

Kneen, M. A.; Annegarn, H. J. *Nucl Instrum Meth B.* 1996, *109*, 209-213.

Lough, G. C.; Schauer, J. J.; Lonneman, W. A.; Allen, M. K. *J Air Waste Manage.* 2005, *55*, 629-646.

Mijić, Z.; Stojić, A.; Perišić, M.; Rajšić, S.; Tasić, M.; Radenković, M.; Joksić, J. *Atmos Environ.* 2010, *44*, 3630-3637.

Mijić, Z.; Stojić, A.; Perišić, M.; Rajšić, S.; Tasić, M. *Chem Ind Chem Eng Q.* 2012, *18*, 623 - 634.

Mijić, Z.; Tasić, M.; Rajšić, S.; Novaković, V. *Atmos Res.* 2009, 92, 420 - 426.

Mijic, Z.; Kuzmanoski, M.; Nicolau, D.; Belegante, L. Proceedings from the 4th WeBIOPATR workshop&conference, October 2-6, Belgrade, Serbia. 2013, 52-59.

Na, K.; Moon, K. C.; Kim, Y. P. *Atmos Environ.* 2005, *39*(30), 5517-5524.

Nordin, E. Z.; Eriksson, A. C.; Roldin, P.; Nilsson, P. T.; Carlsson, J. E.; Kajos, M. K.; Pagels, J. H. *Atmos Chem Phys.* 2013, *13*, 6101-6116.

NSO, (2011). National Statistical Office, report for 2011. [http://webrzs.stat.gov.rs/WebSite/repository/documents/00/00/81/19/Energetski_bilansi_prirodnog_gasa,_nafte_i_derivata_nafte_Republike_Srbije,_2011.PDF]

Olson, D. A.; Norris, G. A.; Seila, R. L.; Landis, M. S.; Vette, A. F. *Atmos Environ.* 2007, *41*, 5673-5683.

Paatero, P. *J Comput Graph Stat.* 1999, *8*, 854-888.

Padhy, P. K.; Varshney, C. K. *Atmos Environ.* 2000, *34*, 577-584.

Perišić, M.; Stojić, A.; Stanišić Stojić, S.; Šoštarić, A.; Mijić, Z.; Rajšić, S. *Air Qual Atmos Hlth.* 2014, 1 - 11.

Rajšić S.; Mijić Z.; Tasić M.; Radenković M.; Joksić J. *Environ Chem Lett.* 2008, *6*, 95-100.

Ran, L.; Zhao, C. S.; Xu, W. Y.; Lu, X. Q.; Han, M.; Lin, W. L.; Chen, L. L. *Atmos Chem Phys.* 2011, *11*, 4657-4667.

Richardson, E. A.; Pearce, J.; Tunstall, H.; Mitchell, R.; Shortt, N. K. *Int J Health Geogr.* 2013, *12*, 34.

Rodriguez-Saona, C. R.; Mescher, M. C.; De Moraes, C. M. *The Role of Volatiles in Plant–Plant Interactions.* In *Long-Distance Systemic Signaling and Communication in Plants*; Springer Berlin Heidelberg, 2013, pp. 393-412.

Romagnoli, E.; Barboni, T.; Santoni, P. A.; Chiaramonti, N. *Nat Hazard Earth Sys.* 2014, *14*(5), 1049-1057.

Saarela, K.; Tirkkonen, T.; Laine-Ylijoki, J.; Jurvelin, J.; Nieuwenhuijsen, M. J.; Jantunen, M. *Atmos Environ.* 2003, *37*, 5563-5575.

Seco, R.; Peñuelas, J.; Filella, I.; Llusià, J.; Molowny-Horas, R.; Schallhart, S.; Hansel, A. *Atmos Chem Phys.* 2011, *11*, 13161-13179.

Singh, H. B.; O'Hara, D.; Herlth, D.; Sachse, W.; Blake, D. R.; Bradshaw, J. D.; Crutzen, P. J. *J Geophys Res Atmos (1984–2012).* 1994, *99*, 1805-1819.

Song, Y.; Dai, W.; Shao, M.; Liu, Y.; Lu, S.; Kuster, W.; Goldan, P. *Environ Pollut.* 2008, *156*(1), 174-183.

Song, Y.; Shao, M.; Liu, Y.; Lu, S.; Kuster, W.; Goldan, P.; Xie, S. *Environ Sci Tech.* 2007, *41*(12), 4348-4353.

Stojić, A.; Stojić, S. S.; Mijić, Z.; Šoštarić, A.; Rajšić, S. *Atmos Environ.* 2015a, *106*, 71-79.

Stojić, A.; Maletić, D.; Stojić, S. S.; Mijić, Z.; Šoštarić, A. *Sci Total Environ.* 2015b, *521*, 19-26.

Taipale, R.; Ruuskanen, T. M.; Rinne, J.; Kajos, M. K.; Hakola, H.; Pohja, T.; Kulmala, M. *Atmos Chem Phys.* 2008, *8*, 6681-6698.

Tasić M.; Đurić-Stanojević B.; Rajšić S.; Mijić Z.; Novaković V. *Acta Chim Slov.* 2006, 53, 401 - 405.

Taylor, K. E. *J Geophys Res Atmos (1984–2012).* 2001, *106*, 7183-7192.

Team, RC (2012). R: A language and environment for statistical computing. [http://cran.case.edu/web/packages/dplR/vignettes/timeseries-dplR.pdf]

Thornhill, D. A.; Williams, A. E.; Onasch, T. B.; Wood, E.; Herndon, S. C.; Kolb, C. E.; Marr, L. C. *Atmos Chem Phys.* 2010, *10*, 3629-3644.

Tomašević M.; Mijić Z.; Aničić M.; Stojić A.; Perišić M.; Kuzmanoski M.; Todorović M.; Rajšić S. *Air Quality Study in Belgrade: Particulate Matter and Volatile Organic Compounds as Threats to Human*, In: *Air Pollution: Sources, Prevention and Health Effects*; Sethi R.; Ed.; Texas A&M Health Science Center (TAMHSC): Bryan, Texas, USA, 2013a.

Tomašević, M.; Antanasijević, D.; Aničić, M.; Deljanin, I.; Perić-Grujić, A.; Ristić, M. *Ecol Indic.* 2013b, *24*, 504-509.

Tomašević, M.; Vukmirović, Z.; Rajšić, S.; Tasić, M.; Stevanović, B. *Chemosphere.* 2005, *61*, 753-760.

Tomašević, M.; Vukmirović, Z.; Rajšić, S.; Tasić, M.; Stevanović, B. *Environ Monit Assess.* 2008, *137*, 393-401.

Uria-Tellaetxe, I.; Carslaw, D. C. *Environ Model Softw.* 2014, *59*, 1-9.

USEPA,USEPA Office of Research and Development. EPA Unmix 6.0 Fundamentals and User guide, 2007.

USEPA. USEPA Office of Research and Development. EPA Positive Matrix Factorization (PMF) 3.0 Fundamentals and User guide, 2008.

Valach, A. C.; Langford, B.; Nemitz, E.; MacKenzie, A. R.; Hewitt, C. N. *Atmos Environ.* 2014, *95*, 456-467.

Vuković, G.; Urošević, M. A.; Razumenić, I.; Goryainova, Z.; Frontasyeva, M.; Tomašević, M.; Popović, A. *Environ Sci Pollut Res.* 2013, *20*, 5461-5470.

Wang, Y. Q. *Meteorol Appl.* 2012, *21*, 360-368.

Watson, J. G.; Chow, J. C.; Fujita, E. M. *Atmos Environ.* 2001, *35*, 1567-1584.

Wilks, D. S. *Statistical methods in the atmospheric sciences* , In: International Geophysics Series Vol. 100; Dmowska R.; Hartmann D.; Rossby H.T. Ed.; Academic Press, Oxford, UK, 2011.

Yuan, B.; Shao, M.; Gouw, J.; Parrish, D. D.; Lu, S.; Wang, M.; Zeng, L.; Zhang, Q.; Song, Y.; Zhang, J.; Hu, M. *J Geophys Res Atmos (1984–2012).* 2012, *117*(D24).

Zheng, J.; Garzón, J. P.; Huertas, M. E.; Zhang, R.; Levy, M.; Ma, Y.; Molina, L. T. *Atmos Environ.* 2013, *70*, 521-531.

Zhu, L.; Huang, X.; Shi, H.; Cai, X.; Song, Y. *Atmos Environ.* 2011, *45*, 594-604.

In: Urban and Built Environments
Editor: Alexis Cohen
ISBN: 978-1-63483-067-6
© 2015 Nova Science Publishers, Inc.

Chapter 5

LOCATION AND ACCESSIBILITY OF NEIGHBORHOODS: TWO KEY FACTORS FOR THE SUSTAINABILITY OF BUILT ENVIRONMENTS

*Anne-Françoise Marique**
Research engineer in architecture and urban planning, University of Liege
Local Environment: Management and Analysis (LEMA),
Chemin des Chevreuils, Liège, Belgium

ABSTRACT

The accessibility and the location of new and existing urban projects are key factors for the sustainability of our built environments. Previous scientific research have shown that both the location of neighborhoods (that is to say their proximity to work places, schools facilities, services etc.) as well as their accessibility (proximity of infrastructures, bus services, etc.) have a huge impact on modal share (especially the ability to use active commuting for daily travel), travelled distances and thus on emissions of greenhouse gas and health of inhabitants. In a view of sustainability, the location of new and planned urban projects should be carefully assessed in order to favor the development of local communities encouraging the use of public transportation and active commuting (foot, bicycle). Local authorities should also be able to assess the potentialities

* Corresponding Author's Email: afmarique@ulg.ac.be.

of different Brownfield and Greenfields on their territories and define where to authorize the development of new urban projects, in order to inscribe them in a concrete vision of sustainability. In the first part of the chapter, we propose a synthetic review of the scientific literature dealing with the relationships between location, accessibility and commuting. Then, we analyze how location and accessibility are taken into account in several existing neighborhood sustainability assessment tools. Based on the ascertainments that sustainable mobility, location and accessibility criteria are given less importance than criteria related to building energy performance or green spaces and, that existing sustainability assessment tools are mainly developed for experts, we propose two frameworks to assess the location and the accessibility of buildings and neighborhoods. These two frameworks are especially designed for a non-expert audience (citizens, local authorities, developers, etc.), within the Walloon (Belgian) context. To conclude the chapter, the reproducibility of our work is discussed and our main findings are summarized.

INTRODUCTION

Transport is a major contributor to the increasing level of energy consumption and greenhouse gas emissions worldwide. In the European Union, the transport sector accounted for 31% of the total final energy consumption in 2005, up from 26% in 1990 (European Commission, 2008). Levels of mobility also continue to rise substantially, and the transport energy demand in 2030 is projected to be 28% higher than in 2005 (European Commission, 2008). Given the levels and growth of energy use in the transport sector, there is an urgent need for measures to control the growth of and reduce the dependence on motorized vehicles and imported oil.

Previous scientific research has shown that both the location of neighborhoods (that is to say their proximity to work places, schools facilities, services etc.) as well as their accessibility (proximity of infrastructures, bus services, etc.) have a huge impact on modal share (especially the ability to use active commuting for daily travel), travelled distances and thus on emissions of greenhouse gas and health of inhabitants.

In a view of sustainability, the location of new and planned urban projects should be carefully assessed in order to favor the development of local communities encouraging the use of public transportation and active commuting (foot, bicycle). Local authorities should also be able to assess the potentialities of different Brownfield and Greenfields on their territories and define where to authorize the development of new urban projects, to inscribe them in a concrete vision of sustainability.

In this chapter, we first present a synthetic state of the art on the relationships between location, accessibility and commuting, at the local scale. Then, we analyze how location and accessibility are taken into account in several existing neighborhood and building sustainability assessment tools. Based on the ascertainments that sustainable mobility, location and accessibility criteria are given less importance than criteria related to building energy performance or green spaces and, that existing tools are mainly expert-oriented, we proposed two specific frameworks to assess the location and the accessibility of buildings and neighborhoods and their impact on energy consumption and green houses gas emissions of a local community. The first one is the first part of a complete handbook that aims at helping local authorities, architects and private developers to build "sustainable neighborhoods." The second one is an online interactive tool that allows a citizen, a household or a local authority to assess and compare energy uses in their house/neighborhood with transportation energy uses. To conclude the chapter, we discuss the reproducibility of our work and summarize our main findings.

SPATIAL PLANNING AND COMMUTING

The relationships between spatial planning (location, accessibility) and commuting (travelled distances, modal shares, energy uses, etc.) have been widely studied in the scientific literature. In the current context of growing interest in environmental issues, favoring active commuting and public transportation and reducing energy consumption in the transport sector, which represents 31% of the overall energy in the European Union (the building sector represents 37%), appears as an important policy target (Maïzia et al., 2009). Existing scientific work dealing with transport consumption is mainly concerned with dense urban areas, focusing on relationships between transport energy consumption and building density, and this work remains undecided on the effects of densification strategies for the reduction of transport consumption. Maïzia et al. (2009) and Steemers (2003) argue that more compact urban forms would significantly reduce energy consumption both in the building and transport sectors. Based on data from 32 big cities located all over the world, Newman and Kenworthy (1989, 1999) have highlighted a strong inverse relationship between urban density and transport consumption, but their work is only valid for certain conditions and is often criticized by other works (Mindali et al., 2004; Owens, 1995) mainly for methodological reasons. Banister (1992) applied the same kind of approach to British cities and highlighted, on the basis of statistical data obtained from a national survey, that transport energy consumption is slightly higher in London than in smaller cities, which refutes Newman and Kenworthy observations. Boarnet and Crane (2001) are also skeptical on the relationship between urban design and transport behaviors. On the basis of several case studies, they estimate that if the use of the soil and the urban form impact transport behaviors, it is through the price of travel (public transport prices are reduced in dense areas). Gordon and Richardson (1997) demonstrated that if fuel prices are included in the analysis, urban density only plays a limited role on energy consumption in transport. Ewing and Cervero (2001), on the basis of a number of case studies, concluded that the impact of urban density on car travel reduction stays marginal. Elasticity is evaluated at -0.05, which means that if the density of a district is multiplied by two, private car commutes are only reduced by 5% because of the rise of congestion. Finally, Breheny (1995) has emphasized minor reductions in transport energy consumption thanks to the compact city model. His experiments show that, even under very strict conditions that are difficult to reproduce, energy used in transport could only be reduced by 10 to 15%. Transport and mobility are crucial in terms of urban form, especially because global oil use has allowed the appearance of sprawling urban forms (Jenks and Burgess, 2002). More recently, Boussauw and Witlox (2009) have developed a commute-energy performance index and tested it for Flanders and the Brussels-capital region in Belgium, including rural and suburban parts of these territories, to investigate the link between spatial structure and energy consumption for home-to-work travels at the regional scale. This method is based on statistical data available at the district scale, taking into account commuting distances, modal shares of non-car travel modes and aspects of infrastructure. This index allows for a better understanding of the energy consumption levels for commuting (home-to-work travels) in cities and less dense areas. Marique et al. (2012) have described a quantitative method that was developed to assess transportation energy consumption at the neighborhood scale. They used this methodology to create a decision-making tool (Marique et al. 2014), to highlight the most efficient strategies needed to promote awareness and to give practical hints on how to reduce energy consumption linked to urban sprawl. Statistical data available at the neighborhood scale and characteristics of cars and public vehicles were used to predict transport needs and assess consumption as far as home-to-work and home-to-school travels are concerned. "Type-profiles" were developed to complete this approach and give an approximation of transport energy consumption related to leisure and commercial purposes. Dujardin et al. (2014) have shown that spatial planning is a huge driver of change in mobility, through their analysis of the impact of territorial structures upon energy uses for transportation and travelled distances. They highlighted that much more progressive approach than the compact city model should be implemented, at the local scale, to take into account the numerous local specificities of urban, suburban and rural settlements. They showed that the diversity of uses, at the local scale, has a major impact on travelled distances and active commuting, as already often highlighted in the literature.

The more specific field of children commuting and especially school commuting, has been investigated, more recently. Since the first works by Hillman et al. (1990), research on child and youth mobility has largely focused on the declining use of active commuting (cycling and walking) and children's independent mobility (the opportunity for children to move freely in the environment without an accompanying adult) in the West, as well as the increasing prevalence of physical inactivity and obesity among children and youth (Buliung et al., 2012; Gilbert and O'Brien, 2005). Research studies have examined correlations between personal, family, social and environmental factors and children's active/ independent commuting to and from school, mainly focusing on health and obesity. The main conclusions of these studies are that the likelihood of walking or cycling to school decreases as travel distance increases, but it is also correlated with the safety of routes and family time constraints (Dellinger and Staunton, 2002; Ewing et al., 2004; Faulkner et al., 2010; McDonald and Aalborg, 2009; McMillan, 2006; Mitra et al., 2010; Sjolie and Thuen, 2002; Timperio et al., 2006; Trapp et al., 2011; Yeung et al., 2008; Ziviani et al., 2004). Previous research has also identified parents' concerns about the absence of adult supervision, the age of the child and fears related to children being assaulted by strangers as primary barriers to active commuting to school. In this respect, gender affects school travel. Boys seem to be more likely to walk and especially to cycle to school (McDonald, 2012; McMillan et al., 2006). An emerging research topic not only concerns the qualitative work on parents' perceptions but also addresses children's perceptions of school travel and the built environment (Faulkner et al., 2010; Fusco et al., 2012; Line et al., 2010). Children are now considered ''active agents'' who are able to express their own needs about school travel (Fusco et al., 2012). Finally, recent studies have highlighted the importance of taking into account both social and environmental issues (Feng et al., 2010; Gilbert and O'Brien, 2005; McMillan 2005, 2006; Saelens and Handy, 2008) and they have attempted (in this regard) to indicate how cities can be changed to improve children's independent mobility. These strategies can be grouped into three broad categories: changing urban form and transport, changing the design of local neighborhoods and changing social values (Tranter and Sharpe, 2012). Although exceptions exist (e.g., Hillman et al. (1990) in England, Sjolie and Thuen (2002) in Norway and Susilo and Waygood (2012) in Japan), most research on children's mobility has been conducted in North America. Relatively less is known about children's mobility in Europe, although European regions and built environments have several significant differences from the countries discussed above. In Belgium, Marique et al. (2013) analyzed the energy consumption, travel distances and mode choices for school commuting based on two decennial surveys. Their results highlighted the following: (1) mobility behaviors have evolved drastically over the past decades for school commuting, an evolution that cannot be entirely related to land-use variables, (2) the energy consumption for school commuting is strongly dependent upon the school level, and (3) the links between land-use patterns and energy consumption for school commuting are different than those highlighted within the literature between urban forms and home-to-work commutes. The concentration of secondary schools and tertiary institutions in urban centers induces higher energy consumption rates, whereas the decentralization of nursery and primary schools across the entire territory leads to very low local energy consumption and increased walking and cycling. Boussaux et al. (2014) assessed the spatial distribution of primary school to examine the extent to which home-to-school trip length is influenced by the spatial distribution of school sites. They namely showed that school accessibility varies considerably, depending on the structure of the concerned settlements.

Location in Existing Sustainability Tools

Despite the extensive literature on the huge relationships between spatial planning (location, accessibility), and transportation (energy uses, modal shares, greenhouses gas emissions), they are rarely taken into account when the sustainability of urban and suburban structures is studied, even if sharp fluctuations in oil prices and reduction efforts in greenhouse gases emissions play an important role in current discussions and policies. Even new districts that set themselves up as "eco" or "sustainable" are sometimes built far from city centers and are not necessarily very sound from an ecological point of view because of higher transport energy consumption (Harmaajärvi, 2000; Marique and Reiter, 2011).

Politicians, stakeholders and even citizens are now aware of the issue of energy consumption in buildings, namely through the passing of the European Energy Performance of Buildings Directive (EPBD), its adaptation to the Member States laws and its recent recast (2010) that imposes all new buildings built by 2020 to be "nearly zero energy" buildings; however, efforts and regulations to control transport needs, location of new buildings and urban projects and consumption related to transportation are much more limited, not to say inexistent in current European regulations.

Several buildings/neighborhoods sustainability assessment tools have been developed during the last years and have become widespread. Many communities, mainly in the developed world, are utilizing these tools to measure their success in approaching sustainable development goals (Sharifi and Murayama, 2013). As neighborhood sustainability assessment (NSA) tools can be considered as the latest generation of impact assessment tools (Sharifi and Murayama, 2014), we will especially focus on this scale, in the following of this paper.

In their tentative to assess and reduce the impact of urban projects on the environment, most of these NSA tools propose to consider contextual issues, such as accessibility, traffic impact of the proposed development, and its location, inspired by the smart growth principles (Sharifi and Murayama, 2013). In their extensive analysis of three well-known NSA tools (LEED-ND in the United States, BREEAM Communities in the UK and CASBEE-UD in Japan), Sharify and Murayama (2014) highlighted that LEED-ND puts a heavy focus on location, connectivity and walkability. Amongst the criteria used in LEED-ND, the "smart location and linkage" package comprises indicators related to school proximity, preferred locations, reduced automobiles dependence and housing and jobs proximity. In LEED-ND, the maximum points achievable for the theme dedicated to location and site selection reaches 18%, against 5% in BREEAM Communities and only 3% in CASBEE-UD (Sharifi and Murayama, 2013). These authors argue that location is a high priority in LEED-ND because, compared to others countries, sprawl is a more severe concern in the United States. BREEAM Communities put a particular focus on transportation by including, for example, criterion related to the development of transport assessment/statement and travel plans and proposing maximum distance from each building entrance to a compliant transport node. In its check-list to favor the sustainable development of neighborhood, the

HQE2R tool proposes a target dedicated to the limitation of urban sprawl and the management of spaces and another one that aims at favoring active commuting. Sharifi and Murayama (2013) also noted that most of NSA tools are biased towards criteria for resources and environment and pattern and design. Criteria related to water, energy, resources conservation and design elements dominate less tangible socio-economic issues.

The SIPRIUS tool, developed by Rey (2012) and applied in a project underway in Neufchatel by Laprise et al. (2015), is an indicator system for the assessment of sustainability integrated into the project dynamics of regeneration of disused areas. It proposes several criteria and indicators related to mobility and accessibility such as the assessment of the quality of service in public transportation, the average distances to schools, kindergarten, commercial zones, recreational facilities, etc. Limit and target values are proposed to inscribe a project in a sustainability perspective.

In a more quantitative approach, Marique and Reiter (2014) have developed a simplified framework to assess the feasibility of the "zero-energy" objective at the neighborhood scale. This framework takes into account, in an holistic approach, building energy uses (for heating, lighting, cooking, etc.), the local production of renewable energies and transportation energy uses of inhabitants for daily mobility (home-to-work, home-to-school, etc.). Energy uses can be quantified and compared. In the same vein, Riera-Perez and Rey (2013) studied the influence of infrastructure, facilities and services on daily mobility, in a Swiss case study. They used a method developed by the Swiss Society of Engineers and Architects (2011) to estimate the energy consumption related to mobility. In the following of Boussaux and Witlox (2011), they highlighted that a holistic vision, reducing distances and promoting public transportation is required. Favoring more sustainable behavior of inhabitants is also highlighted as an important target, in the previous papers.

All these tools and frameworks are interesting in themselves and in their specific context. They feed and improve the debates about smart growth and sustainability of built environments but they present two main limitations: (1) the assessment is time and money-consuming and (2) these tools are mainly expert-oriented, excluding citizens and local actors; both of which make their concrete utilization by citizens and local actors quite difficult.

Two Frameworks Dedicated to Non-Experts

In this context, based on general and specific trends highlighted in the literature dedicated to spatial planning, transportation, location and accessibility, and at the request of the Regional administration, two frameworks specifically dedicated to a non-specialized audience were developed in Wallonia (Belgium). These framework must allow to overpass the two major limitations highlighted in the previous section: (1) a neighborhood assessment is time and money-consuming and (2) existing tools are expert-oriented, excluding citizens and local actors although they are the first actors that can concretely act, in practice, to reduce transportation energy uses (through spatial planning and good location choices but also through behavioral changes).

A Handbook for Sustainable Neighborhood

In a view of sustainability, the location of new and planned urban projects must be carefully assessed in order to favor the use of public transportation and active commuting (foot, bicycle), at the local / neighborhood scale. Local authorities must also be able to assess the potentialities of different Brownfield and Greenfields on their territories and define where to authorize the development of new urban projects, to inscribe them in a concrete vision of sustainability.

The first framework is thus dedicated to local authorities, architects and private developers and aims at helping them to take into account location and accessibility when developing or retrofitting an urban project, at the neighborhood scale. To be widely used and quick to apply, this framework is simple, accurate and easy to learn and to use.

The framework presented in this section is only one part of the complete "handbook for sustainable neighborhood," developed by Teller et al. (2014), at the request of the Minister of urban planning. Others parts are respectively dedicated to density, the use of resources (energy, materials, etc.), natural and green spaces, equipments and social diversity and participation.

This framework is built upon four main criteria of accessibility and location, which are:

- Train services: this criterion takes into account the distance from the neighborhood to the train station and the type of station (regional or local). The distance to a regional train station must be lower than 1.500 meters or the distance to a local train station must be lower than 1.000 meters.

- Bus services: this criterion takes into account the number of buses per day in the surrounding of the neighborhood (the assessment takes place in an area of 700 meters around the boundaries of the neighborhood). Target values are adapted according to the location of the neighborhood: minimum 34 buses per day in poles and city-centers, min 20 in villages and rural cores.
- Diversity of functions: the neighborhood must be located in a mixed-use environment to favor active commuting and reduce travelled distance. The target values are: at least 15 different functions (alimentary shops, bakery, post office, etc.) in an area of 700 meters around the boundaries of the neighborhood in poles and city centers; at least 5 in villages and rural cores. Moreover, at least, three different categories of functions must be represented (amongst supermarket, shops, leisure, services, public services).
- Proximity of school facilities: there is a least one school in the surroundings of the neighborhood.

These four criteria are firstly assessed separately (Figure 1). They are major criteria in the overall evaluation provided in the handbook: 25 sustainability criteria are proposed. To deserve the "sustainable neighborhood" label in Wallonia, 20 out of the 25 criteria must be respected. Amongst those criteria, at least two out of the three ones must be respected to ensure that the neighborhood is located in an existing core with good public transportation services and opportunities to reduce travelled distances, through a good diversity of functions, at the local scale.

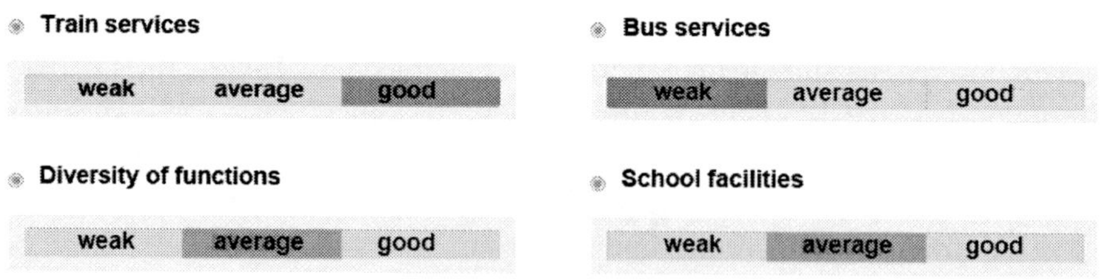

Figure 1. The four accessibility criteria are assess separately, which allow to easily highlight strengths and weaknesses of the considered location.

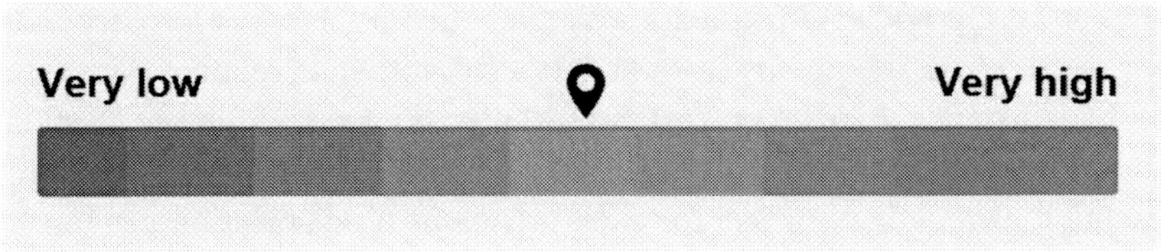

Figure 2. The four criteria are combined to build a composite indicator of sustainability. This indicator allows, for example, to compare the accessibility of different locations.

These four criteria can then be combined to build a composite indicator of accessibility that is useful to objectively compare the location of different neighborhoods or sites. Three values (good, average or bad) can be applied at each criterion, based on the results of the assessment. The combination of all possible choices gives 81 possibilities. These 81 possibilities are then ranked in order to assess the overall accessibility of the neighborhood, as highlighted on figure 2.

This handbook was politically validated and is available since February 2014. This handbook is currently concretely used in several urban projects under progress, as well as for the assessment of existing and planned neighborhoods.

An Online Interactive Tool to Assess Energy Consumption for Daily Mobility

This second framework is part of an overall online interactive tool to assess energy consumption in buildings (heating, cooling, appliances, etc.) and for daily mobility (Marique et al., 2014).

As far as transportation energy uses are concerned, a quantitative method was developed to assess energy consumption for daily mobility. The complete method and data set were presented in detail in two previous papers addressing the investigation of the relationships between the urban form, the location of the activities in the territory and the transportation energy consumption for home-to-work and home-to-school travels (Marique et al., 2013a and 2013b). Three indices were derived from this general method to be used in the interactive tool. The energy performance index (expressed in kWh/travel person) for a defined unit (a city, a neighborhood, a household or a person) represents the mean energy consumption for home-to-destination travels for one person living within the considered unit. This index takes into account the distances travelled, the means of transport used and their relative consumption rates. The distance index (in km) represents the mean home-to-destination distance travelled (one way) by one person. The modal share index (in %) represents the frequency of use for each means of transportation per territorial unit. The annual energy efficiency of the travels can finally be expressed in kWh per year to enable a comparison between the energy consumption for daily mobility and the energy consumption in the residential building sector (heating, appliances, electricity, etc.). The input data used in the method provided in the online interactive tool are obtained from either national census results or, preferably, directly from the users

The online interactive tool provides three types of evaluation: a rapid evaluation, a detailed evaluation and a neighborhood evaluation. They are based on the same methodology and are respectively dedicated to an individual user without any knowledge about energy, an individual user with limited knowledge and a local authority or a private developer. In the framework of this chapter, only the detailed evaluation (the most complete one for an individual user or a household) is presented.

This detailed evaluation allows an individual user (or household) to assess the amounts of energy consumption for daily mobility (numerous purposes of travels and chained trips can be taken into account). The user is first asked to precisely define the composition of the user's household (number of adults and of children) and the transport habits of each member of the household, in an online questionnaire. For each type of travel or chained trip, it is necessary to specify the modes, the travelled distances, the numbers of travels per week and per year. Detailed questions about the consumption of the vehicle, the type of fuel and car sharing are asked if the mode is the car. The energy consumption for daily mobility is calculated via the method developed by Marique et al. (2013a) and according to the travel characteristics of the user and results are proposed to the user in a very intuitive way. These results, related to transportation energy uses, can also be compared with building energy uses of the users, in an overall view of sustainability (Figure 3). Finally, the online interactive tool provides several improvements to reduce travelled distances and transportation energy uses, adapted to the personal habits of the user. The potential energy savings related to each improvement are quantified.

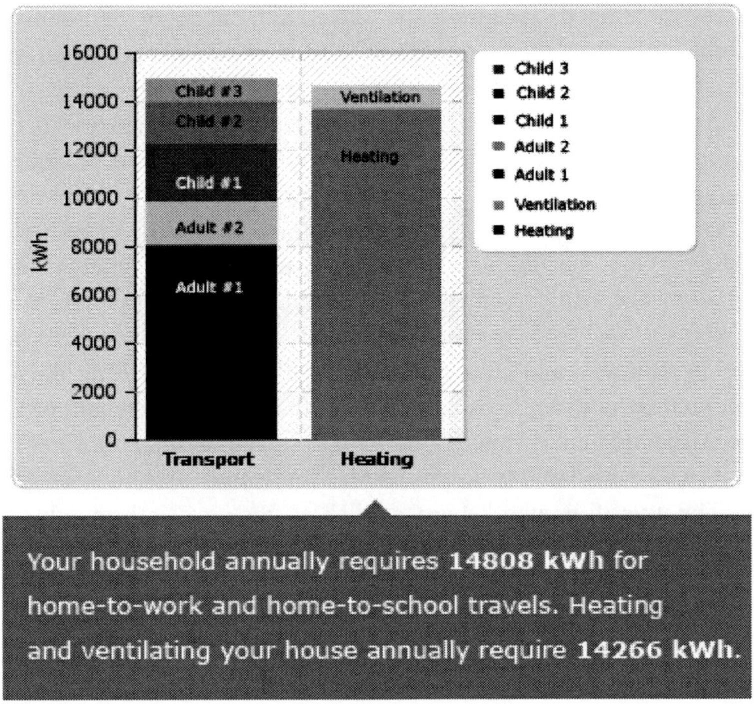

Figure 3. Example of results (the comparison between transportation and building energy uses).

This online interactive tool (building + transportation) is available for free on www.safe-energie.be. This version of the tool is available since 2013 and is concretely used by citizens and local authorities (400 visits each month). During the development of the online interactive tool, the research team organized several test sessions and comments from the users were quite positive. In the following of this first tool, an updated version, comprising new developments, such as the production of renewable energies on site, was recently developed. The final version of this updated tool is available since 2015, on www.solen-energie.be.

CONCLUSION

Location and accessibility are important factors for the sustainability of our built environments. They have a huge impact on travelled distances, modal share and transportation energy uses. Several sustainability assessment tools have been developed recently and take into account, amongst others sustainability criteria, these factors in the assessment of urban projects. However, they are time and money consuming and mainly expert-oriented, which make them difficult to use by local actors and citizens. To overpass these limitations, two frameworks, specifically designed for a non-specialized audience, were developed in Wallonia (Belgium). They allow easily and quickly assessing and comparing the accessibility of Greenfields and Brownfields. One of them also allows to quantify energy uses for transportation, according to the location of activities (work places, schools, houses, etc.) on the territory. These frameworks were developed to be easily usable by local actors and citizens. They were developed within the Walloon (Belgian) context with local data. However, the methodology and criteria that have been used are general enough to be reproduced in others different regions and territories, but adapting targets values and input data to each particular context. The two frameworks are available since several months and are currently concretely used by local actors and citizens.

ACKNOWLEDGMENTS

The two frameworks presented in this paper were respectively developed under two research program: "A handbook for sustainable development" and "SOlutions for Low Energy Neighboudhood," both funded by the Walloon Region of Belgium (SPW-DGO4).

REFERENCES

Banister, D. Energy use, transport and settlement patterns. In: Breheny M, editor. *Sustainable Development and Urban Form.* London: Pion Ltd; 1992; 160-181.

Boarnet, M., Crane, R. Travel by design. *The influence of urban form on travel. New*-York: Oxford University Press; 2001.

Boussauw, K., Witlox, F. *Introducing a commute-energy performance index for Flanders.* Transportation Research Part A, 2009 43, 580-91.

Boussaux, K., van Meeteren, M., Witlox, F. Short trips and central places: The home-school distances in the Flemish primary education system (Belgium). *Applied Geography,* 2014 53, 311-322.

Breheny, M. The compact city and transport energy consumption. *Transactions of the Institute of British geographers,* 1995 20(1), 81-101.

Buliung, R., Sultana, S., Faulkner, G. Guest editorial: special section on child and youth mobility – current research and nascent themes. *Journal of Transport Geography,* 2012 20, 31–33.

Dellinger, A.M., Staunton, C.E. Barriers to children walking and bicycling to school – United States, 1999. *Morbidity and Mortality Weekly,* 2002 51 (32), 701–704.

Dujardin, S., Marique, A.-F., Teller, J. Spatial Planning as a driver of change in mobility and residential energy consumption. *Energy & Buildings,* 2014 68, 779–785.

European Commission. European energy and transport – trends to 2030. Luxemburg: European commission; 2008.

Ewing, R., Schroeer, W., Greene, W. School location and student travel: analysis of factors affecting mode choice. *Transportation Research Record,* 2004 1895, 55–63.

Faulkner, G.E.J., Richichi, V., Buliung, R.N., Fusco, C., Moola, F. What's quickest and easiest? Parental decision making about school trip mode. *International Journal of Behavioral Nutrition and Physical Activity,* 2010 7, 1–11.

Feng, J., Glass, T.A., Curriero, F.C., Steward, W.F., Schwartz, B.S. The built environment and obesity: a systematic review of the epidemiologic evidence. *Health & Place,* 2010 16, 175–190.

Fusco, C., Moola, F., Faulkner, G., Buliung, R., Richichi, V. Toward an understanding of children's perceptions of their transport geographies: (non) active school travel and visual representations of the built environment. *Journal of Transport Geography,* 2012 20, 62–70.

Gilbert, R., O'Brien, C. *Child- and Youth-Friendly Land-Use and Transport Planning Guidelines.* The Toronto: Centre for Sustainable Transportation; 2005.

Gordon P, Richardson H W. Are compact cities a desirable planning goal? *Journal of the American Planning Association,* 1997 63(1), 95-106.

Harmaajärvi I. EcoBalance model for assessing sustainability in residential areas and relevant case studies in Finland. *Environmental Impact Assessment Review,* 2000 20, 373-80.

Hillman, M., Adams, J., Whitelegg, J. One false move… *A Study of Children's Independent Mobility.* London: Policy Studies Institute; 1990.

Jenks, M., Burgess, R., editors. *Compact cities: sustainable urban forms for developing countries.* London: Spon Press; 2002.

Laprise, M., Lufkin, S., Rey, E. An indicator system for the assessment of sustainability integrated into the project dynamics of regeneration of disused urban area. *Building and Environment,* 2015 86, 29-38.

Line, T., Chatterjee, K., Lyons, G. The travel behaviour intentions of young people in the context of climate change. *Journal of Transport Geography,* 2010 18, 238–246.

Maïzia, M., Sèze, C., Berge, S., Teller, J., Reiter, S., Ménard, R. Energy requirements of characteristic urban blocks. In: *Proceedings of the CISBAT 2009 International Scientific Conference on Renewables in a Changing Climate,* From Nano to Urban scale, Lausanne; 2009.

Marique, A.-F., Reiter, S. Towards more sustainable neighbourhoods: are good practices reproducible and extensible? In Bodard, M., Evrard, A., (Eds.), Proceedings of International Conference PLEA 2011 : *Architecture & Sustainable Development (*pp. 27-32). Presses Universitaires de Louvain; 2011.

Marique, A.-F., Reiter, S. A method for evaluating transport energy consumption in suburban areas. *Environmental Impact Assessment Review,* 2012 33, 1-6.

Marique, A.-F., Dujardin, S., Teller, J., Reiter, S. School commuting: the relationship between energy consumption and urban form. *Journal of Transport Geography,* 2013a 26, 1-11.

Marique A.-F., Dujardin S., Teller J., Reiter S., Urban sprawl, commuting and travel energy consumption, *Proceedings of the Institution of Civil Engineers, Energy,* 2013b 166, 1–13.

Marique, A.-F., de Meester, T., De Herde, A., Reiter, S. An online interactive tool to assess energy consumption in residential buildings and for daily mobility. *Energy and Buildings,* 2014 78C, 50-58.

Marique, A.-F., Reiter, S. A simplified framework to assess the feasibility of zero-energy at the neighbourhood / community scale. *Energy and Buildings,* 2014 82, 114-122.

McDonald, N.C. Getting to school: the impact of free transit on low income and minority students. *Annual Meeting of the Transportation Research Board.* Washington, DC; 2005.

McDonald, N.C., Aalborg, A.E. Why parents drive children to school. *Journal of the American Planning Association,* 2009 75 (3), 331–342.

McMillan, T.E. Urban form and a child's trip to school: the current literature and a framework for future research. *Journal of Planning Literature,* 2005 19, 440–456.

McMillan, T.E. The relative influence of urban form on a child's travel mode to school. *Transportation Research Part A,* 2006 41, 69–79.

McMillan, T., Day, K., Boarnet, M., Alfonzo, M., Anderson, C. Johnny walks to school – does Jane? Sex differences in children's active travel to school. Children, *Youth and Environments,* 2006 16 (1), 75–89.

Mindali, O., Raveh, A., Salomon, I. Urban density and energy consumption: a new look at old statistics. *Transportation Research Part A: Policy and Practice,* 2004 38, 143–162.

Mitra, R., Buliung, R.N., Roorda, M.J. The built environment and school travel mode choice in Toronto, Canada. *Transportation Research Record,* 2010 2156, 2150– 2159.

Newman, P., Kenworthy, J. *Cities and Automobile Dependence: A Sourcebook.* Victoria: Gower, Aldershot and Brookfield; 1989.

Newman, P., Kenworthy, J. *Sustainability and Cities: Overcoming Automobile Dependence.* Washington, DC: Island Press; 1999.

Owens, S. A response to Michael Breheny. *Transactions of the Institute of British geographers,* 1995 20, 381-86.

Rey, E. *Régénération des friches urbaines et développement durable : vers une évaluation intégrée à la dynamique de projet.* Louvain-La-Neuve : Presse Universitaires de Louvain ; 2012.

Riera-Perez, M., Rey, E. Estimated reduction of energy consumption related to mobility in urban renewal projects. *Case study in Lausanne, Switzerland. Proceedings of PLEA* 201; 2013.

Saelens, B.E., Handy, L. Built environment correlates of walking: a review. *Medicine & Science on Sports & Exercise*, 2009 40, 550–566.

Sharifi, A., Murayama, A. A critical review of seven selected neighborhood sustainability assessment tools. *Environmental Impact Assessment Review,* 2013 38, 73-87.

Sharifi, A., Murayama, A. Neighborhood sustainability assessment in action: Cross-evaluation of three assessment systems and their cases from the US, the UK, and Japan. *Building and Environment,* 2014 72, 243-248.

Sjolie, A.N., Thuen, F. School journeys and leisure activities in rural and urban adolescents in Norway. *Health Promotion International,* 2002 17 (1), 21–30.

Steemers K. Energy and the city: density, buildings and transport. *Energy and Buildings,* 2003 35(1), 3-14.

Susilo, Y.O., Waygood, E.O.D. A long term analysis of the mechanisms underlying children's activity-travel engagements in the Osaka metropolitan area. *Journal of Transport Geography,* 2012 20, 41–50.

Swiss Society of Engineers and Architects. *La voie SIA vers l'efficacité énergétique, compléments et exemples relatives au cahier technique SIA* 2040. Documentation D/SI1 0236/F.Zürich : SIA ; 2011.

Teller, J., Marique, A.-F., Loiseau, V., Delbar, C., Godard, F. *Le référentiel quartier durable.* Namur: SPW-DGO4 ; 2014.

Timperio, A., Ball, K., Salmon, J., Roberts, R., Giles-Corti, B., Simmons, D., Bauer, L.A., Crawford, D. Personal, family, social and environmental correlates of active commuting to school. *American Journal of Preventive Medicine,* 2006 30, 45–51.

Tranter, P., Sharpe, S.. Disney-Pixar to the rescue: harnessing positive affect for enhancing children's active mobility. *Journal of Transport Geography,* 2012 20, 34–40.

Trapp, G.S.A., Giles-Corti, B., Christian, H.E., Bulsara, M., Timperio, A.F., McCormack, G.R., Villanueva, K.P. Increasing children's physical activity: individual, social and environmental factors associated with walking to and from school. *Health Education and Behaviour,* 2011 39 (2), 172–182.

Yeung, J.,Wearing, S., Hills, A.P. Child transport practices and perceived barriers in active commuting to school. *Transportation Research Part A,* 2008 42, 895–900.

Ziviani, J., Scott, J., Wadley, D. Walking to school: incidental physical activity in the daily occupations of Australian children. *Occupational Therapy International,* 2004 11 (1), 1–11.

In: Urban and Built Environments
Editor: Alexis Cohen
ISBN: 978-1-63483-067-6
© 2015 Nova Science Publishers, Inc.

Chapter 6

EFFICIENCY OF PLANNING AND DESIGN TOOLS TOWARDS ECO-FRIENDLY HIGH-DENSITY COMMUNITY

***Yu-Hsin Tsai*[1,*], *Meng Ting Wu*[2] *and Yuwen Huang*[3]**
[1]Department of Land Economics, National Chengchi University, Taipei, Taiwan
[2]Taiwan Railway Association, Taiwan
[3]State-owned Enterprise Commission, Ministry of Economic Affairs, Taiwan

ABSTRACT

High-density development in a city or community has been increasingly promoted in urban planning arena to achieve sustainable, ecological goals through impact on transportation behaviour and less per capita land consumption. Such high-density development to achieve region-wide ecological purpose, however, may have a negative ecological impact at the community level in terms of worsened natural light, wind, and the space needed for humans, animals, and vegetation to live healthily, and in turn affects the acceptance of high-density policy. One of the solutions to mitigating the negative impact may lie in the spatial arrangement of building mass and outdoor space since they determines the fundamental spatial characteristics in the community; however, few studies have been carried out in this area. This paper aims to present an

[*] Corresponding author: E-mail: yhsin@nccu.edu.tw.

inventory of urban planning and design tools for arranging building mass and open space, and to examine their impacts on some eco-friendly indexes. A real case of urban renewal site is applied for simulation analysis to assess the impacts of the planning tools, while in order to evaluate the general impact of each planning tool a hypothetical city is developed, where other factors can be controlled for. The results suggest that lowering the building coverage rate and increasing height-distance ratio are the two most efficient tools for improving the overall quality of eco-cities for the selected indexes. The efficiency of the tools in improving individual aspects of an eco-city, such as pedestrian-friendly environment, green space accessibility, adaptive capacity of CO_2 absorption, and animal biodiversity are also gauged. These research findings can help improve the eco-city planning or design guidelines, which may work more effectively citywide in developing eco-city than the more labor-intensive case-based simulation review process, popular in many cities.

Keywords: high-density, sustainability, eco-city, urban planning, urban design

In the pursuit of sustainable or eco-city development, the high-density built environment policy of smart growth, compact city, and transit-oriented development (TOD) has been regarded as an essential tool in urban planning (Costello, 2005). High-density residential planning is likely to degrade the residential quality of the physical environment, with opposing market and planning forces; therefore sensitive planning and design criteria and tools are necessary. Moreover, high-density development may harm ecological conditions such as natural light, wind, and space, which are vital to the health of the primary stakeholders of residential areas, including humans, animals, and vegetation. The issue is complicated by the necessity to consider multiple ecological conditions as a whole, which unfortunately remains unresolved. Furthermore, the choice of planning and design tools available to planners and their level of influence in shaping eco-neighbourhoods have been inadequately studied in the past. This issue is particularly important for city core residential areas, where high density makes the urban environment less healthy and enjoyable for the stakeholders.

More specifically, the challenge to the physical quality of high-density neighbourhoods may be mitigated by better urban planning and through the spatial arrangement of building mass or outdoor space. Planning and design tools include building coverage rate (BCR) and building setback (SB); however, their level of impact on ecological characteristics for the primary stakeholders has barely been examined. Hence, the purpose of this research is threefold: to establish an inventory of urban planning and design tools that control the built environment of a neighbourhood, to develop eco-city aspects and indexes in the urban planning realm rather than in the fields of architecture and environmental protection, and to evaluate the impacts of planning and design tools on these eco-city indexes. The report includes a literature review on the eco-city, planning and design tools, eco-city indexes, and the efficiency of planning and design tools in shaping eco-neighbourhoods. It also includes research methods for evaluating efficiency, results of simulation analyses, and conclusions and policy implications.

1. The Eco-City and Urban Planning and Design

According to Downton (2009), there is no standard definition of the eco-city. The concept is more goal-led (outcome-oriented) than spatial-form-oriented (input-oriented) and can be classified into three levels: less damage to the overall ecosystem in terms of resource consumption and pollution, a healthy built environment for humans, and the preservation and/or restoration of natural habitats for vegetation and wildlife. The eco-city has been advocated by environmentalists, architects, and urban planners, among others. In the arena of urban planning, eco-city implementation tools are mainly focused on land-use planning and include high–density development (Kasanko et al., 2006), mixed–use development (Cervero et al., 2011), jobs-housing balance (Cervero, 1996a), and pedestrian-friendly environments (Cervero & Sullivan, 2011) as well as increased green space, such as parks, greenbelts, and green sidewalks, to expand the ecological landscape. Some of these large-scale spatial planning tools are intended to increase accessibility and enrich green transportation alternatives.

At the other end of the spatial-scale spectrum, green building design and materials are used by architects to increase energy efficiency, reduce carbon emission, and mitigate the urban heat island (UHI) effect. Mid-scale neighbourhood and block planning and design tools have not been studied as much, but they appear to have the potential to achieve the goals of eco-city design from different perspectives. Theoretically, the spatial arrangement of building mass or space in a neighbourhood determines the location, height, and/or rough shape, neglecting the details, of buildings, which, in turn, may affect ecological characteristics such as natural lighting (Martin & March, 1973; Yang et al., 2007), wind, the UHI effect, the conservation of space or natural habitat for planting or wildlife, green space accessibility, and the suitability of the environment for pedestrians.

1.1. Urban Planning and Design Tools: Shaping the Residential Built Environment

Past research, planning and design guidelines or ordinances, and case studies provide a range of tools for the arrangement of building mass at the neighbourhood level (Tsai, 2010). For example, density distribution and open space in the land-use plan determine the large-scale or citywide building mass arrangement, whereas the spatial arrangement of building mass at the neighbourhood and block levels can be controlled by such planning and design tools as building shape (BS), floor area ratio (FAR), BCR, height distance ratio (HDR), side yard width (SW), backyard depth (BD), and SB (Taipei City Government, 2011; Brook McIlroy Planning + Urban Design, 2006).

Each of the block-level tools functions differently in arranging outdoor space or building mass. FAR determines the amount of building mass within a three-dimensional spatial unit, essentially regulating building or population density, or both. BCR determines the extent to which ground space is reserved for such purposes as spatial openness, planting, and drainage. HDR is used to ensure daylight exposure. The stair-shaped SB (SSB) not only works as HDR but also provides more three-dimensional space between buildings along the street. Moving-backwards SB (refer to Table 1 for graphic definition) leaves more street-level space as sidewalk or open space but may result in less space at the back. SW and BD can ensure the minimum distance necessary between adjoining buildings for lighting or fire prevention purposes. The location of outdoor space can be determined by building shape, such as tower, slab and cross.

1.2. Aspects and Indexes of an Eco-City at Neighbourhood Level

This section compiles the aspects and indexes of the ecological neighbourhood relevant to the three primary stakeholders: humans, animals, and vegetation. The first (and least complex) subsection—natural environment and adaptive capacity—concerns the extent to which the natural environment is preserved in high-density urban settings, based on the assumption that the original ecological conditions are best for the native flora and fauna and, naturally, for human communities. Some natural conditions can adapt better than the built environment to environmental conditions and natural disasters; for example, large green, open space resolves the issue of surface runoff and allows the wind to blow off urban heat, and it can also provide abundant vegetation for CO_2 absorption and heat mitigation. The second subsection addresses the potential for human communities to improve the environment by reducing transportation energy consumption and tailpipe pollution at the block level, which is made possible by creating a pedestrian-friendly environment and improving green space accessibility. Unbuilt land in urban areas provides space for vegetation growth, but the natural characteristics of the land may be lost already; hence the unbuilt land may serve more for humans with green environment than for native plants requiring more natural environment to grow in. Finally, the third subsection addresses the concepts and dilemmas involved in preserving animal habitats.

1.2.1. Natural Environment and Adaptive Capacity

1.2.1.1. Daylight Exposure

Sunlight can be impeded by building mass in two ways: building height (Martin & March, 1973) and lack of permeability between buildings (Fisher-Gewirtzman et al., 2005). Conventionally, HDR is adopted as a proxy to manage daylight exposure; however, it has the drawback of considering only the impedance by buildings across the street. Furthermore, sky view factor (SVF), defined as the ratio of sky observed to total sky hemisphere, takes into account sunlight blocked from all directions (Ratti & Richens, 1999). In addition, solar radiation (insolation) is a more delicate indicator, calculated on the basis of direct and diffuse radiation between the winter and summer solstice for a point location or a whole area (ESRI, 2007).

1.2.1.2. Wind Environment

In urban areas, wind speed and direction can be affected by building mass, layout, height, and alignment. Earlier research has addressed the issue of discomfort caused by gusts and strong winds (American Society of Civil Engineers, 1999); however, much of the recent research in this area has focused on weak winds and public health issues such as airborne contagious disease and the UHI effect (Oke, 1982; Wong et al., 2010). Four principal measures have been applied to gauge wind speed in urban settings: field surveys, wind tunnel modeling, numerical models, and morphometric methods (Tsai, 2010). Ground coverage ratio (GCR), defined as the ratio of total built area to total area, is proposed as a substitute for planners on the basis of its simplicity (Ng et al., 2011) for computer simulation for all cases.

1.2.2. Residents

1.2.2.1. Pedestrian-Friendly Environment

Pedestrian-friendly environment, or walkability, has become more significant in achieving sustainability due to its capacity for incubating non-mobile travel behaviour (Lo, 2009). Aside from its quantitative characteristics, its qualitative aspects include accessibility, safety, security, amenities, and diversity of pedestrians and activities. These aspects can be affected by the availability of sidewalk, design of sidewalk network (e.g., routing and intersections) and sidewalk segments (e.g., availability of width and shade) , land use (e.g., density and diversity) and building design (e.g., front yard design) along the sidewalk, and sidewalk buffers to separate pedestrians and traffic (Ramirez et al., 2006).

1.2.2.2. Green Space Accessibility

Parks and community green space can meet the recreational needs of residents. Increased accessibility to green space in a neighbourhood may reduce the distance to amenities and encourage walking and biking. There is a range of accessibility variables, each with a different emphasis and meaning (Tsai, 2010). Euclidian distance, network distance, and travel time emphasise the closest destination as the most significant. On the other hand, the cumulative-opportunities variable emphasises the importance of opportunities within a certain distance. More delicate, but also more difficult, variables include utilities-based (Bertolini & Kapoen, 2005), and level-of-service accessibility indexes (Tsai, 2010), taking both supply and demand into consideration.

1.2.3. Animals

In highly developed urban settings, two distinct dilemmas may be encountered in relation to the preservation or restoration of existing habitats. The first challenge is to develop a migration network within the limited space of a highly built environment using urban planning and design tools. The second dilemma is whether animal welfare is at risk if existing natural habitats are connected to the non-natural landscape, thereby increasing exposure to potential danger from human activities. This dilemma is addressed by two distinct planning ideologies: the biocentric equality policy proposed by Devall

& Sessions (1985), which focuses on reducing exposure issue and the uncertain results of artificially created wildlife habitats, and the policy of "coexistence" of animals and humans in residential areas focusing on the re-appearance of animals in the neighborhood. Ecological condition is often measured by applying the concepts of landscape biology, i.e., the patch-corridor–matrix model (Forman and Godron, 1986), which can represent habitats, migration paths, and buildings in the context of this research.

1.3. Impacts of Planning and Design Tools at the Neighbourhood Level

The spatial arrangement of building mass varies depending on the planning tools used, and this is likely to affect the physical environment and the resulting ecological conditions. The impact or efficiency of each planning and design tool may vary; for example, in a high-density neighbourhood where daylight exposure is a matter of concern, it is unclear which tool is most efficient. Furthermore, each tool is likely to affect various aspects of the physical environment differently; and evaluating their impact is further complicated when multiple aspects need to considered, with potential trade-offs between them. For instance, for the same FAR, lowered BCR can increase planting space but decrease daylight exposure, because building height increases as a result of lowered BCR.

2. Research Methods

To assess the impacts of planning and design tools on eco-city buildings, simulation analyses are conducted for a real case of urban renewal site, and for a hypothetical city in order to evaluate the impact with surrounding built environment controlled for, and for animal biodiversity. The real case and hypothetical city consider environmental sustainability from the aspects of human residents, plants, and the natural environment, but not animals; the simulation analysis for animal biodiversity focuses on developing a migration network for urban animal habitats. The hypothetical city is chosen to complement the real urban renewal case, whose built environment is unique at the site and the surrounding area; this is because the hypothetical case provides a general impact evaluation where the surrounding built environment is the same as the site, excluding the effect of different type of surrounding built environment if selected for simulation analysis. The hypothetical city imitates the high-density city or city core since its natural environments are mostly transformed, and hence the eco-city is of more concern. All the scenarios of the real case and hypothetical city have the same density/FAR to provide an unbiased evaluation baseline. Two analysis techniques are used: the elasticity is calculated to compare the relative efficiency of planning and design tools, and the slope is calculated to identify recommended scenarios. Google SketchUp, AutoCAD, ArcGIS, and Excel are used in the analysis.

2.1. Real Case and Hypothetical City

The real case is a residential/commercial urban renewal site located in Taipei, Taiwan, currently with two four-story apartment buildings (Figure 1) covering a 42 m * 35.9 m site, accommodating 32 residential units. Some developers proposed urban renewal design plans in the past few years waiting for agreement from the residents.

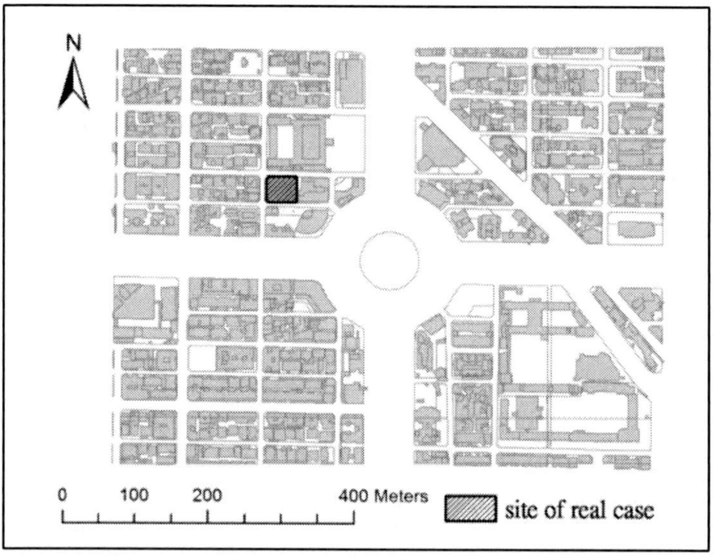

Figure 1. Real case for eco-city simulation analysis- Taipei, Taiwan.

The BCR and FAR for the current buildings are 60% and 240%, respectively. While the new BCR and FAR after renewal will be changed to up to 50% and 456%, respectively. A boundary of 400 m radius is set for conducting the simulation analysis, which provides enough space for index calculations at the one hand, while is not too large to compute on the other hand.

The hypothetical city designed for the eco-city simulation can be broken down into land use, road, and park systems. Its land use layout is composed of 50 m * 50 m blocks. Each block group comprises 4 * 4 blocks, 4 * 4 block groups constitute one neighbourhood, 2.5 * 2.5 neighbourhoods constitute a community, and 2 * 2 communities constitute the 5.07 km * 5.07 km hypothetical mini-city (Figure 2), which is surrounded by natural habitats. Within each block lie four uniform buildings of the highest development density (300% FAR).

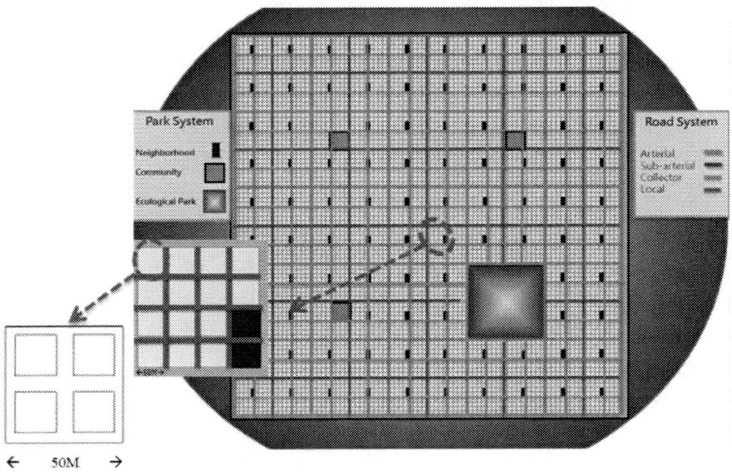

Figure 2. Base map of hypothetical city for general eco-city simulation analysis.

These are adopted from Taipei city, which is a type IV residential area (Taipei City Government, 2011). The tower–shaped buildings have 50% BCR and 3 m storey height and face either north or south. This hypothetical city is large enough to form building clusters at its centre in order to gauge ecological indexes with little difference or distortion relative to larger cities. The land-use plan does not intentionally distinguish residential and commercial areas, etc. For simplicity, it distinguishes only between buildings and open space. Land use is served by a four-tier road hierarchy comprising 8 m local streets, 20 m collectors, 30 m sub-arterial streets, and arterials (Figure 2). All but the 8 m local streets have sidewalks to simulate the situation in Taipei city.

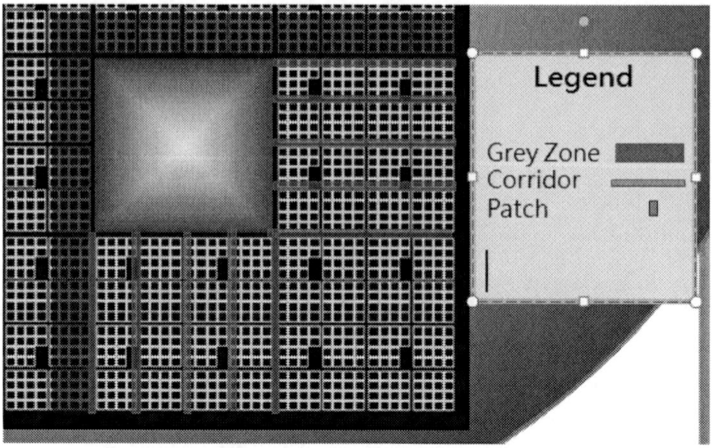

Figure 3. Hypothetical community for animal-biodiversity simulation analysis.

The park system, unlike other infrastructure, is incorporated into the model due to its significance ecologically and in evaluating green space accessibility. It comprises neighbourhood, community, and ecological parks wherein natural habitats remain, and it is located in the southeast quadrant (Figure 2). This quadrant is modeled in more detail for the animal-biodiversity simulation (Figure 3). The ecological park is confined by a 200-meter-wide "grey buffer" of no plants around the north and west sides and linked by some linear north-south and east-west-oriented corridors on the outskirts to minimise potential exposure to human activities. Although the network is animal-centric, the assessment of the impacts of planning tools will also apply, to a certain extent, to the policy of coexistence of animals and humans as long as the network is rewired accordingly, such as linked to other residential areas.

2.2. Urban Planning and Design Tools and Scenarios

Assuming that other non-space factors such as building materials and colour are the same, six space-based urban planning and design tools are pooled together to develop various scenarios for baseline building shape (tower): BCR, HDR, SB, SW, BD, and street-corner SB (SCSB) (Table 1). In addition to the tower shape (BS1), seven other building shapes are established: enclosed courtyard (BS2), slab (BS3), U-shaped (BS4), convex-shaped (BS5), cross-shaped (BS6), X-shaped (BS7), and reverse U-shaped (BS8). For each of the planning and design tools and building shapes, a set of scenarios is developed at fixed intervals within the largest reasonable range of variation. If possible, five scenarios are developed to calculate the efficiency/elasticity of each tool (refer to the evaluation techniques below). The descriptions and two- and three-dimensional concept graphics for the scenarios are presented in Table 1.

2.3. Eco-City Indexes: Two Types of Simulations

Two sets of eco-city indexes are developed for the real case and general eco-city, and animal-biodiversity simulation analyses, respectively. The former considers the stakeholders—residents, plants, and the natural environment and its adaptive capacity—and the latter focuses on animal migration network building. These indexes are developed to achieve a range of goals from the perspectives of the different stakeholders and also to characterise the built environment.

Firstly, the eco-city can be approached by reducing travel energy consumption and tailpipe pollution by creating a pedestrian-friendly environment (for trips of all purposes) and green space accessibility (for local recreational trips), which can also meet residents' demands for a more natural environment (Table 2). The mean of these two indexes, equally weighted, is taken to quantify the overall pedestrian-friendly environment. The reason for the equal weighting is that the two indexes may differ under various conditions in terms of people, city, and time, and an element of arbitrary judgment is unavoidable but hard to be reach universal, objective agreement. By the same token, this spirit of second best solution of equal weighting is applied to the other aspects discussed below and to the overall physical eco-city index. Secondly, more open space for plants to grow is an eco-city goal. Thirdly, urban settings that remain closer to the original level of daylight exposure and wind environment without buildings are considered more natural for both humans and plants. Insolation, or daylight exposure, is calculated as the weighted mean magnitude of nine numeric points and four alphabetic points (Figure 4). It is measured, using ArcGIS, for residents walking on the street and at the front and back of the building on the ground level of the central block to represent the whole city. Furthermore, green space is used to quantify the city's adaptive capacity in terms of CO_2 absorption and heat mitigation. Finally, patch, T-link, and T-Gap-related indexes are used to characterise patches and corridors for the animal-biodiversity simulation (Table 3). All the variables are standardised within the range 0–1.

Table 1. Scenarios of urban planning and design tools and building shape

Urban Planning/Design Tools and BS: Scenario Descriptions[1]	2-D Concept Graphics: Scenarios (Overlaid)	3-D Concept Graphics	
1. BCR Scenarios: BCR30, 35, 40, 45 ...75(%) Buildings shrink inward or expand outwards around building central point as BCR changes.		BCR30 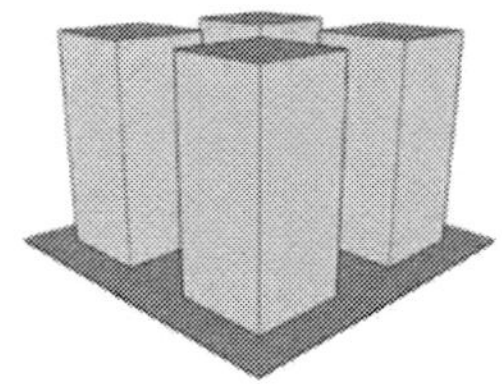	BCR75 
2. HDR 2-1. HDR-S: Floor area from *side* The distance between buildings across the street (south and north sides) is fixed, with height varying for different scenarios at 0.25 intervals. The floor area for the increased stories is shifted from the (right) side of the building. Scenarios: HDR-S1.25, 1.50 … 2.25		HDR-S1.25 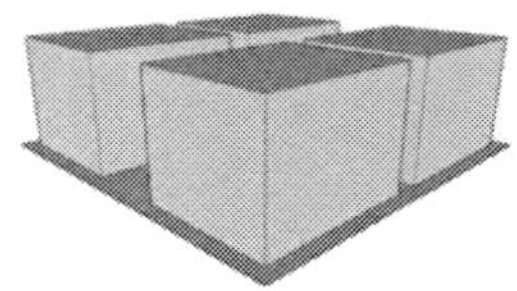	HDR-S2.25 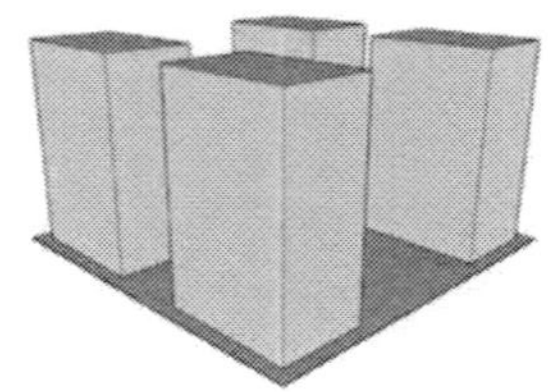

Urban Planning/Design Tools and BS: Scenario Descriptions[1]	2-D Concept Graphics: Scenarios (Overlaid)	3-D Concept Graphics	
2-2. HDR-B: Floor area from *back* The distance is fixed with height varying for different scenarios at 0.25 intervals. The floor area for the increased stories is shifted from the back of the building. Scenarios: HDR-B1.25, 1.50 … 2.25	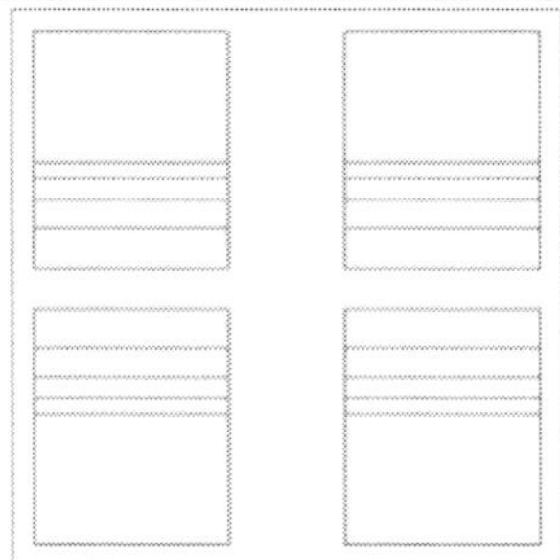	HDR-B1.25 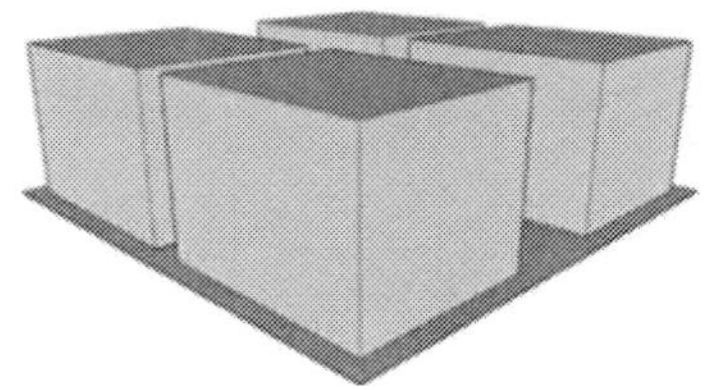	HDR-B2.25 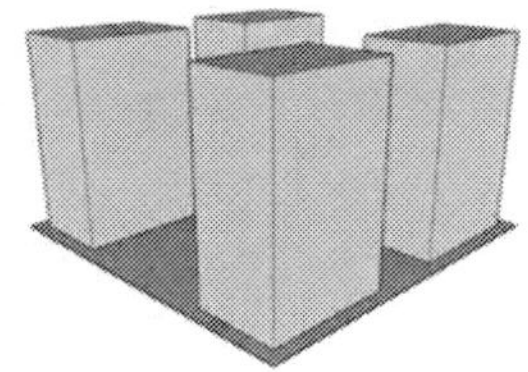
3. SB 3-1. SB-B: Moving *backwards* The whole building moves backwards towards the block centre at 1.5 m intervals for various scenarios. Scenarios: SB-B0, 1.5, 3, 4.5, 6 (m)		SB-B0 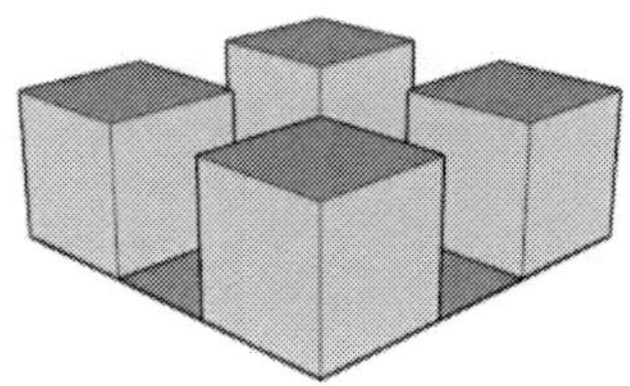	SB-B6 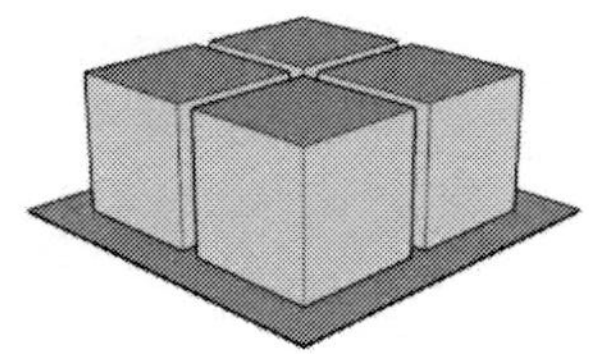

Table 1. (Continued)

Urban Planning/Design Tools and BS: Scenario Descriptions[1]	2-D Concept Graphics: Scenarios (Overlaid)	3-D Concept Graphics	
3-2. SB-T: Floor area to *top* The whole building shrinks towards the south-east corner of the building at 1.5 m intervals for various scenarios, and the cutoff floor area is shifted to the top of the building. Scenarios: SB-T0, 1.5, 3, 4.5, 6 (m)	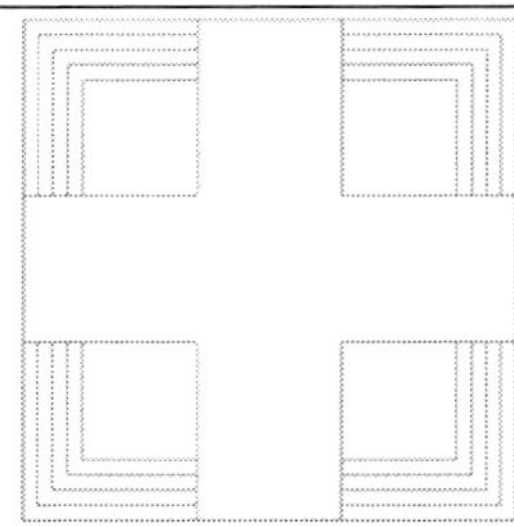	SB-T0 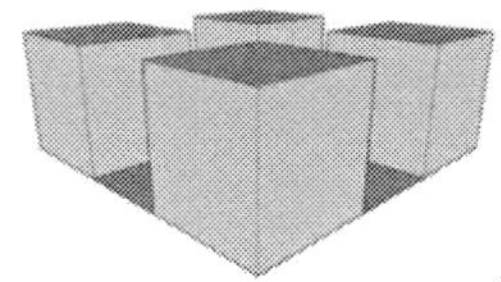	SB-T6 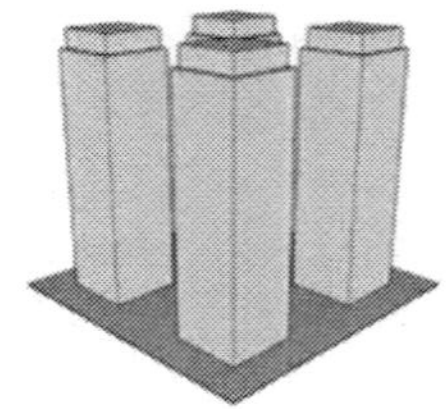
4. SW 4-1. SW-B: Floor area to *back* The side yard width is increased at 1 m intervals, and the cutoff floor area is shifted to the back of the building. Scenarios: SW-B1.5, 2.5, 3.5, 4.5, 5.5(m)	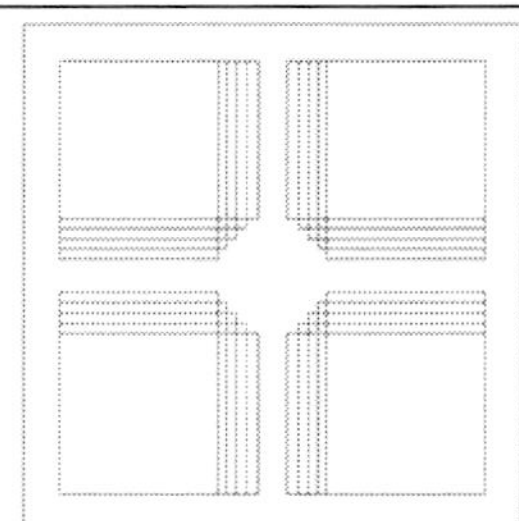	SW-B1.5 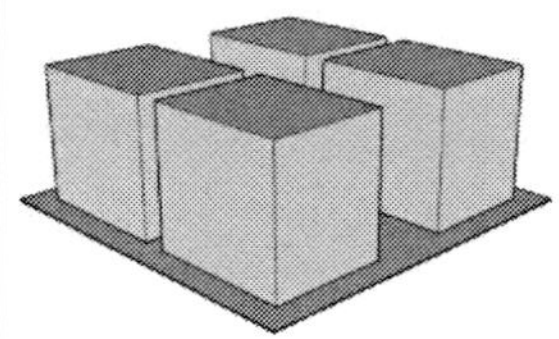	SW-B5.5 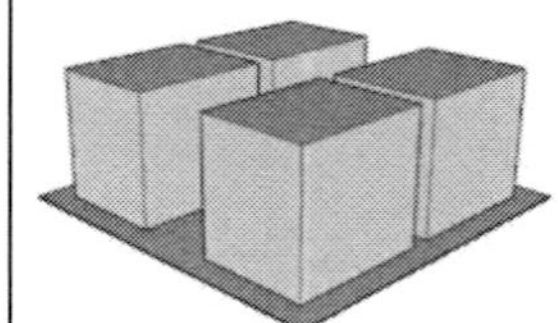

Urban Planning/Design Tools and BS: Scenario Descriptions[1]	**2-D Concept Graphics: Scenarios (Overlaid)**	**3-D Concept Graphics**	
4-2. SW-T: Floor area to *top* The side yard width is increased at 1 m intervals, and the cutoff floor area is shifted to the top of the building. Scenarios: SW-T1.5, 2.5, 3.5, 4.5, 5.5 (m)	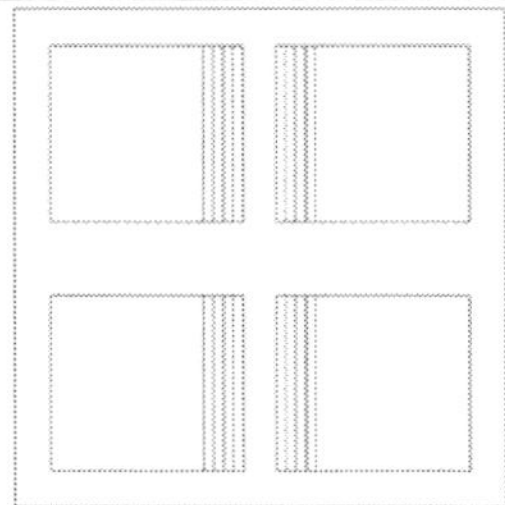	SW-T1.5 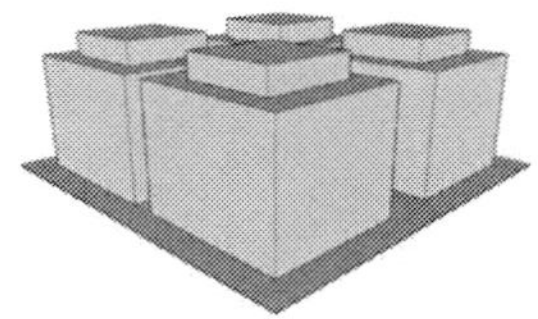	SW-T5.5 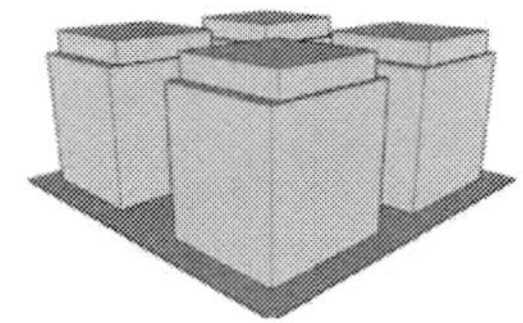
5. BD 5-1. BD-S: Floor area to *side* The backyard depth is increased at 1 m intervals, and the cutoff floor area is shifted to the right side of the building. Scenarios: BD-S1.5, 2.5, 3.5, 4.5, 5.5 (m)	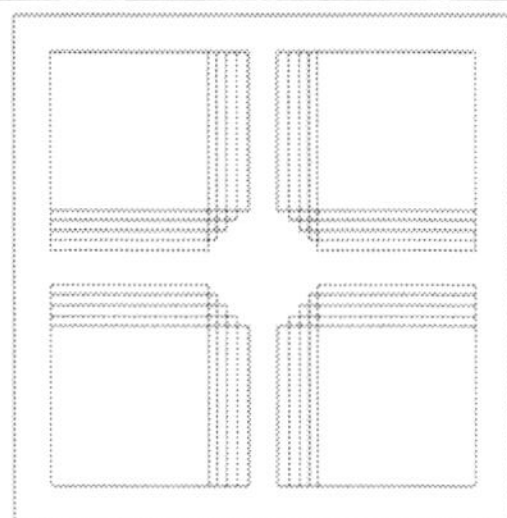	BD-S1.5 	BD-S5.5 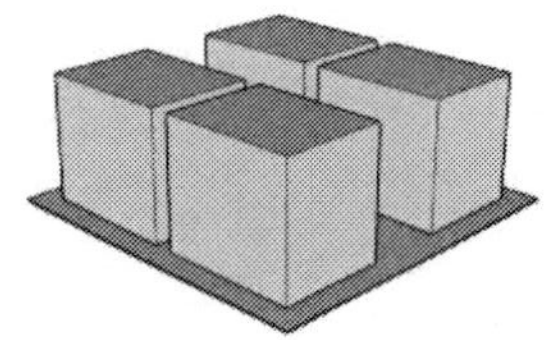

Table 1. (Continued)

Urban Planning/Design Tools and BS: Scenario Descriptions[1]	2-D Concept Graphics: Scenarios (Overlaid)	3-D Concept Graphics	
5-2 BD-T: Floor area to *top* The backyard depth is increased at 1 m intervals, and the cutoff floor area is shifted to the top of the building. Scenarios: BD-T1.5, 2.5, 3.5, 4.5, 5.5 (m)		BD-T1.5 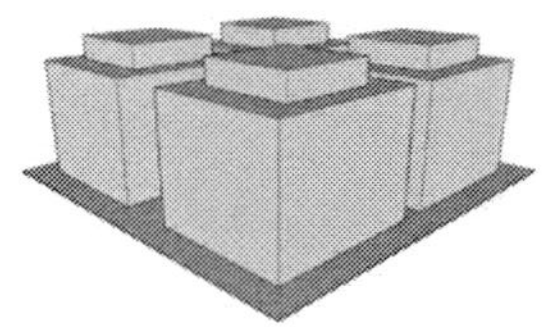	BD-T5.5 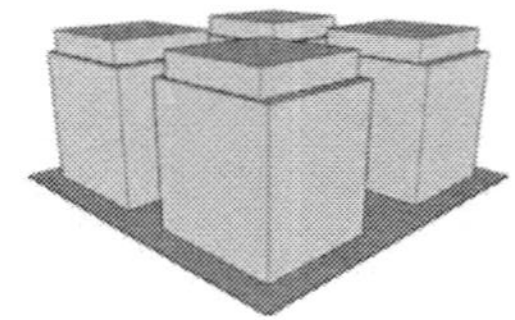
6. SCSB 6-1. SCSB-H-1F: Horizontal setback on 1F The building corner at the intersection is set back from the ground level and 0, 2, 4, and 6 meters above for various scenarios, and the cutoff floor area is shifted to the top of the building. This can also be called horizontal setback. Scenarios: SCSB-H-1F0M, 2M, 4M, 6M	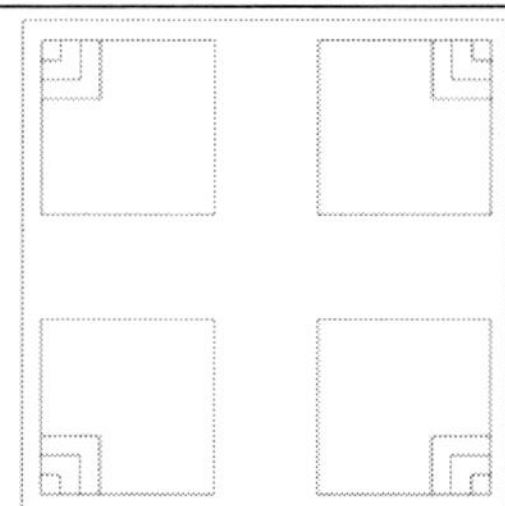	SCSB-H-1F0M 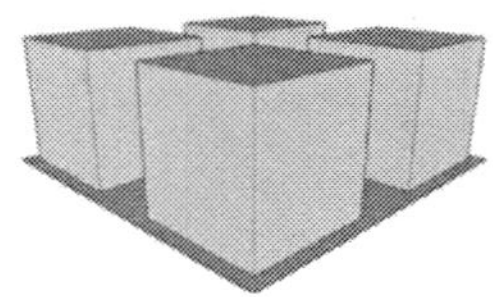	SCSB-H-1F4M 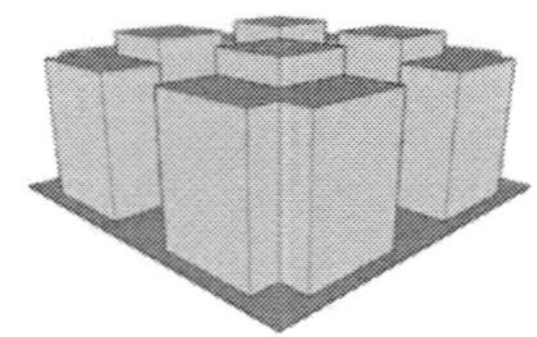

Urban Planning/Design Tools and BS: Scenario Descriptions[1]	2-D Concept Graphics: Scenarios (Overlaid)	3-D Concept Graphics	
6-2. SCSB-V-4M: Vertical 4 m setbacks The building corner at the intersection is set back for 4 meters starting from 3F, 2F, and 1F for various scenarios, and the cutoff floor area is shifted to the top of the building. This can also be called vertical setback. Scenarios: SCSB- V-4M-None, 3F, 2F, 1F	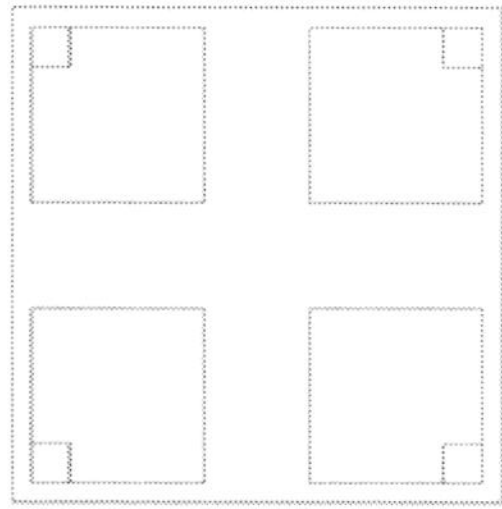	SCSB-V-4M1F 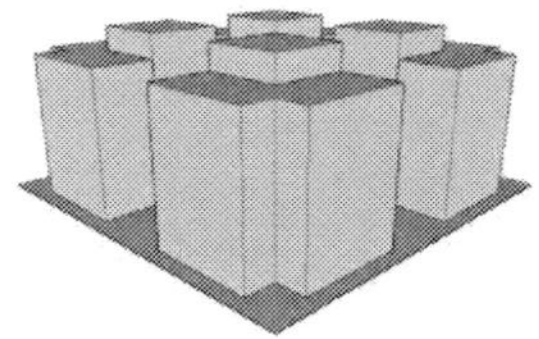	SCSB-V-4M2F 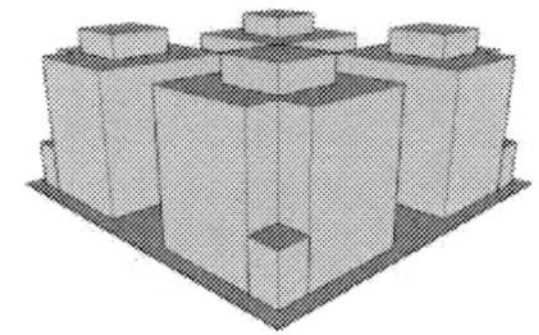
7. BS BS2. Enclosed courtyard The front yard depths for various scenarios are 1.5, 2, 2.5, and 3 m; the cutoff floor area is moved into the courtyard. Scenarios: BS2-1.5, 2, 2.5, 3 (m)		BS2-1.5 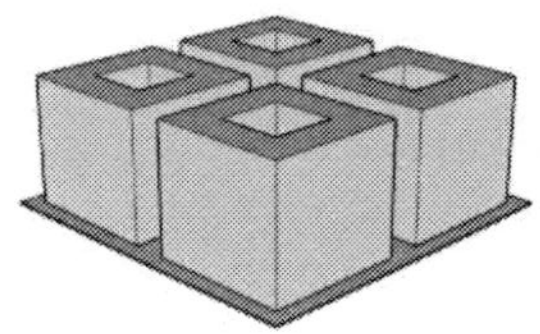	BS2-2.5 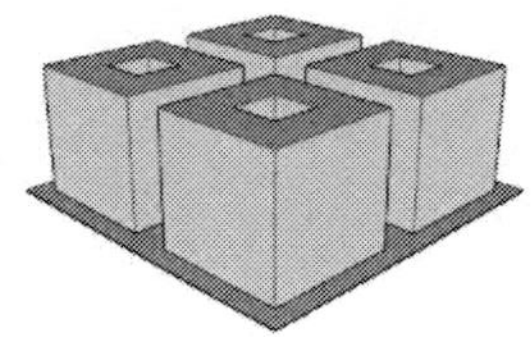

Table 1. (Continued)

Urban Planning/Design Tools and BS: Scenario Descriptions[1]	2-D Concept Graphics: Scenarios (Overlaid)	3-D Concept Graphics	
BS3. Slab The side yard widths for various scenarios are 1 and 2 m. Scenarios: BS3-1, 2 (m)		BS3-1 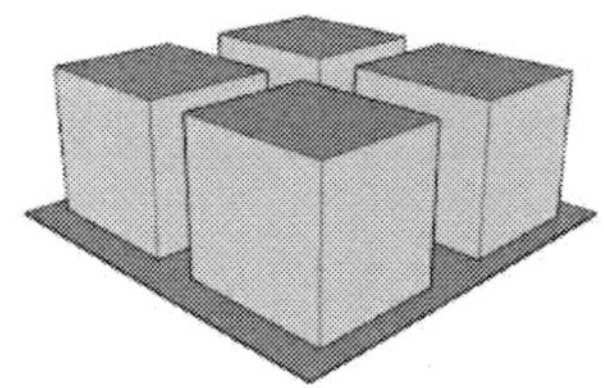	BS3-2 
BS4. U-Shaped The depths of the pits of the U-shaped buildings vary at 1 m intervals. Scenarios: BS4-1, 2, 3 (m)	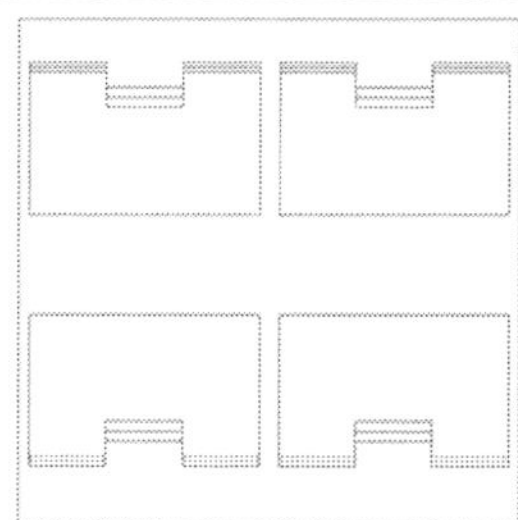	BS4-1 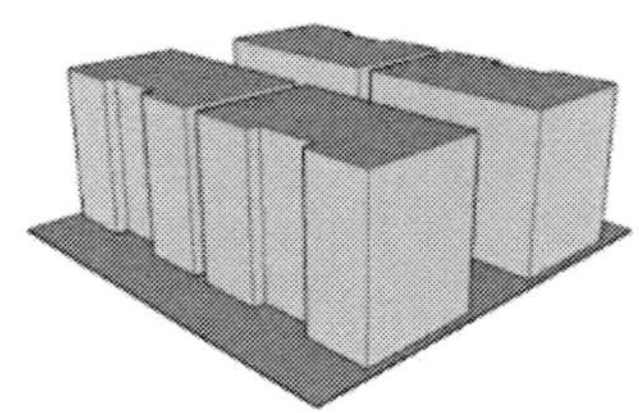	BS4-3 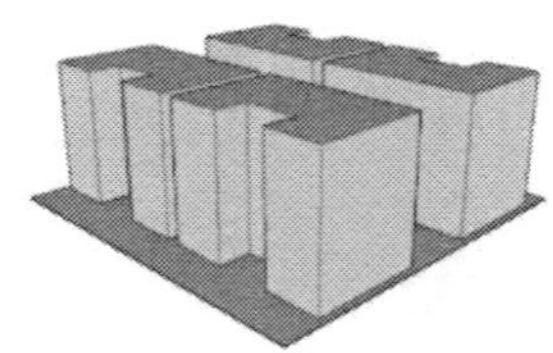

Urban Planning/Design Tools and BS: Scenario Descriptions[1]	2-D Concept Graphics: Scenarios (Overlaid)	3-D Concept Graphics	
BS5. Convex-shaped The lengths of the front bulges of the convex-shaped buildings vary at 1 m intervals. Scenarios: BS5-1, 2, 3 (m)	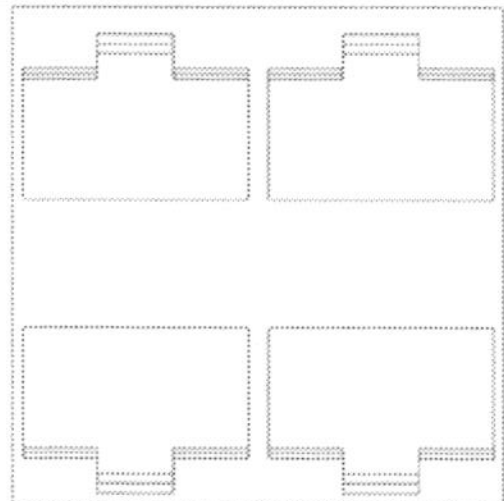	BS5-1 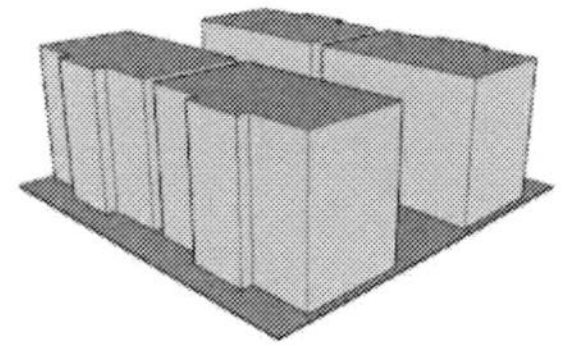	BS5-3 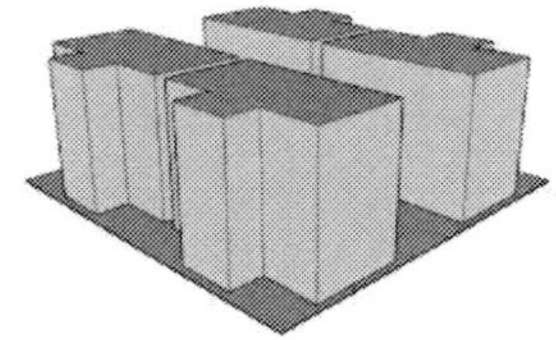
BS6. Cross-shaped The lengths of the four arms of the cross-shaped buildings vary at 0.5 m intervals. Scenarios: BS6-2, 2.5, 3 (m)	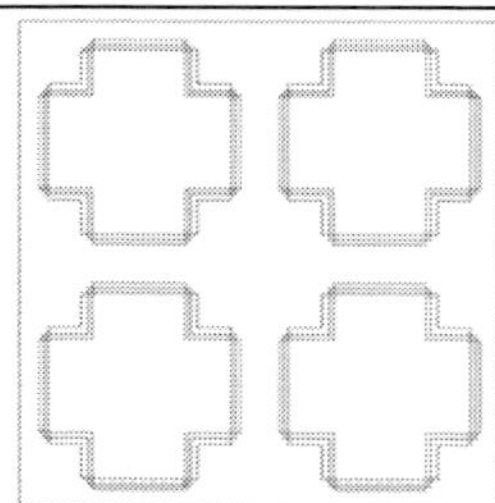	BS6-2 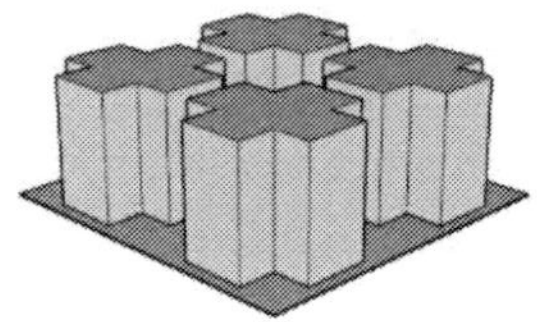	BS6-3 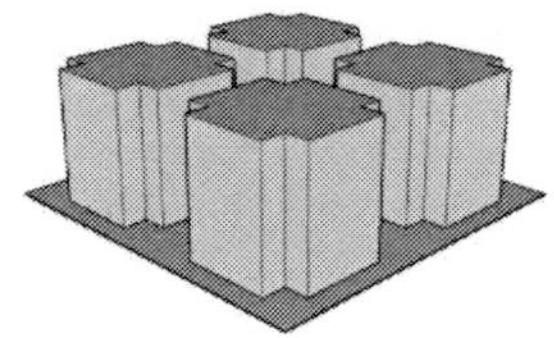
BS7. X-shaped The lengths of the leg parts of the buildings vary at 0.5 m intervals. Scenarios: BS7-2, 2.5, 3 (m)	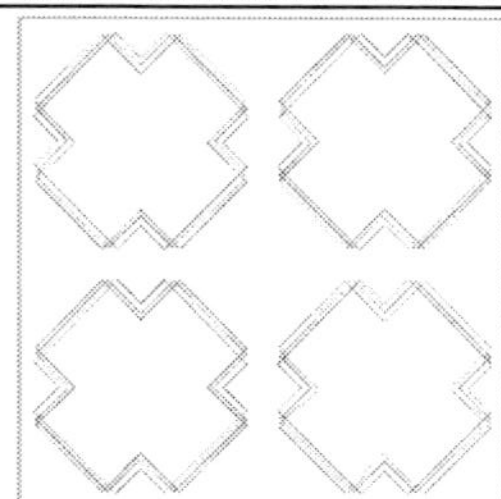	BS7-2 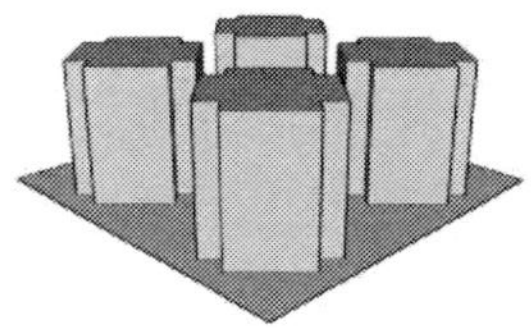	BS7-3 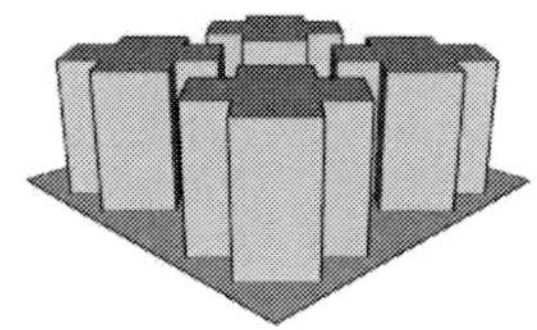

Table 1. (Continued)

Urban Planning/Design Tools and BS: Scenario Descriptions[1]	2-D Concept Graphics: Scenarios (Overlaid)	3-D Concept Graphics	
BS8. Reverse U-shaped The depths of the dent parts of the buildings vary at 1 m intervals. Scenarios: BS8-1, 2, 3 (m)	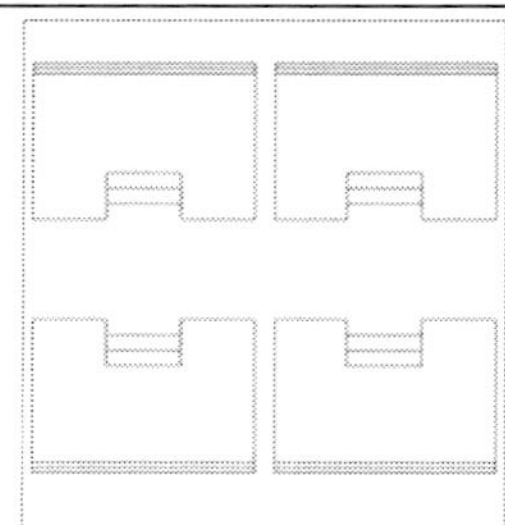	BS8-1 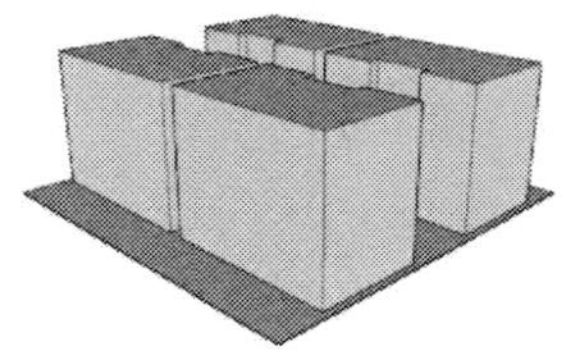	BS8-3 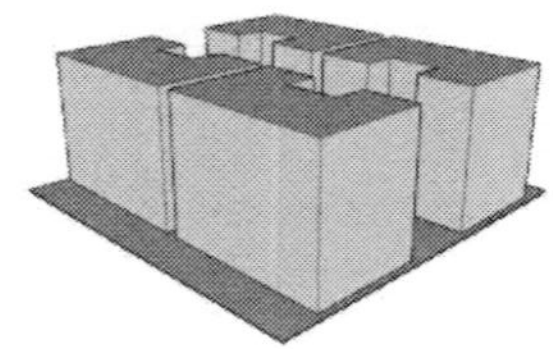

1. The upper-left building in the block is used as the example in the scenario description.

Table 2. Eco-city indexes for general eco-city simulation, by stakeholder and by impact aspect

Stakeholder	Impact Aspect	Index
Residents	Pedestrian-Friendly Environment	$\text{Sidewalk Percentage} = \frac{\text{Sidewalk Length}^{*}}{\text{Street Length}^{**}}$
		$\text{Green Sidewalk Percentage} = \frac{\text{Green Sidewalk Length}^{***}}{\text{Street Length}^{**}}$
		Overall Index=Mean of above Two Indexes
	Green Space Accessibility	$\text{Green Space Percentage} = \frac{\text{Green Open Space Land Area}^{****}}{\text{Total Land Area}}$ [within 5-minute Walking Distance (400m)]
		Green Space Distance (m) = Distance to The Closest Park, Open Space or Yard Larger Than 500 Square Meters***** $\text{Standardised Green Space Distance Index} = \frac{\text{Green Space Distance}}{\text{Largest Green Space Distance of All Scenarios}}$
		Overall Index=Mean of above Two Indexes
Plants	Open Space for Plants	$\text{Green Coverage Ratio (GCR)} = \frac{\text{Total Leaf Area}}{\text{Total Land Area}}$
Natural Environment and Adaptive Capacity	CO_2 Absorption, Heat Mitigation	$\text{Green Coverage Ratio (GCR)} = \frac{\text{Total Leaf Area}}{\text{Total Land Area}}$
	Wind Environment	$1 - \text{Ground Coverage Ratio} = 1 - \frac{\text{Total Built Area}}{\text{Total Land Area}}$
	Daylight Exposure	$\text{Insolation Percentage}^{*****} = \frac{\text{Insolation (Watts/m}^2)}{\text{Insolation without Buildings}}$
Overall Eco-City Index		Overall Index=Mean of above Six Impact Indexes

* Sidewalk length is the total length of the street segments where the width between the kerb and building façade is larger than 1.5 m for installation of minimal sidewalk.

** The street length excludes the intersection segments, where sidewalk cannot be implemented.

*** Green sidewalk length is the total length of the street segments where the width between kerb and front wall of the building is larger than 3 m, allowing trees or turf to be grown on the sidewalk.

**** Green open space is defined as those parks, open spaces, or yards larger than 500 m^2, which is the most popular park size (500–1000 m^2) and the bottom 15.4-percentile of parks of the Taipei city land-use plan.

***** The location setting is Taipei city, and the time period is the second half of the year 2011, between the first day of summer and the last day of fall.

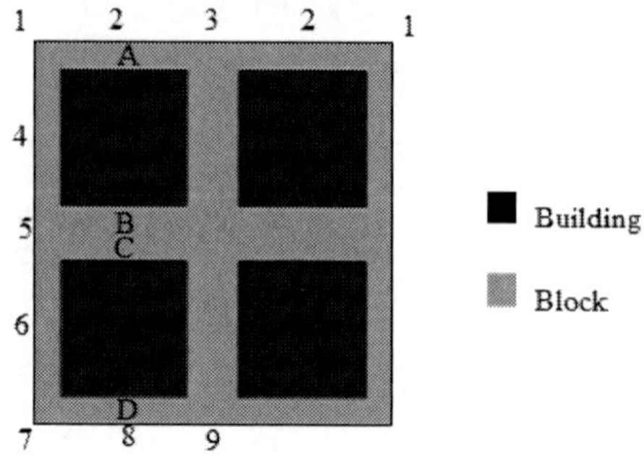

Figure 4. Sampling points for measuring insolation.

Table 3. Eco-city indexes for animal-biodiversity simulation, by aspect

Stakeholder	Aspect	Index
Animals (Primarily birds and insects)	Patches	$\text{Patch Number Percentage} = \frac{\text{Number of Open Space } (\geq 500\text{m}^2)}{\text{Largest Number of Open Space}}$
		$\text{Patch Land Area Percentage} = \frac{\text{Total Area of Open Space } (\geq 500\text{m}^2)}{\text{Total Land Aaea}}$
		Overall index = Mean of above two indexes
	Corridor	$\text{T Link Number Percentage} = \frac{\text{Number of Green Corridors}}{\text{Largest Number of Green Corridors}}$
		$\text{T Link Land Area Percentage} = \frac{\text{Total Area of Green Corridors}}{\text{Total Land Aaea of Corridors}}$
		Overall index = Mean of above two indexes
	Quality of Corridor	$1\text{-(T-Gap)} = \frac{\text{Total Length of Green Segments of Corridors}}{\text{Total Length of Green Corridors}}$
Overall Animal-Biodiversity Eco-City Index		Mean of above three aspect indexes

2.4. Evaluation Techniques: Elasticity and Slope

To evaluate the efficiency of urban planning and design tools in improving eco-city indexes, the slope and mid-point elasticity are used. The slope is primarily used to identify scenarios where the eco-city index rises or falls, representing the most significant change. Mid-point elasticity is used to cross-examine the relative efficiencies of various planning tools by taking advantage of its unit-free capacity. The elasticity is adopted and modified to measure the percentage change in an eco-city indicator as a response to the percentage change of the planning and design tool (Equation 1). The mid-point/arc elasticity is also used to calculate the average elasticity since scenarios for each planning tool cover a certain range.

$$E_{EC,Tool} = \frac{\partial PL/PL}{\partial Tool/Tool}$$ Equation 1

where $E_{EC,\,Tool}$ is the urban planning/design tool elasticity of the eco-city index

EC is the mid-point magnitude of the eco-city index

∂EC is the change in magnitude of the eco-city index

Tool is the mid-point magnitude of the urban planning/design tool

∂Tool is the change in magnitude of the urban planning/design tool

3. Impacts of Urban Planning and Design Tools on Eco-City Indexes

This section first reveals the impacts of urban planning and design tools from the simulation analysis of the hypothetical city which provides a more general evaluation since its surrounding built environment is generalized rather than a unique type in any other real case if selected. Then, due to the complex redundant calculation, results and analysis, only selected, significant results from simulation analysis of the real case are presented.

3.1. Impacts of BCR

The values of all eight indexes for the hypothetical city are presented for the eco-city aspects in Table 4-1. The statistics for pedestrian-friendly environment indexes indicate that when the BCR falls to 55%, the overall index value jumps significantly and then increases marginally, suggesting a maximum BCR of 55% in this regard. The values for sidewalk percentage show that there is enough room (the space between kerb and building wall is wider than 1.5 m (U.S. Department of Justice, 1999; Taiwan Ministry of The Interior, 2004)) to build sidewalks on both sides of the streets for all BCR scenarios.[8] However, the sidewalk is only wide enough (3 m) to be greened

[8] Sidewalk should generally be built on the side of a road rather than on private property. However, due to the non-existence of sidewalk on most of the narrow streets in traditional

when the BCR is reduced to 55% or below. For the same reason, the accumulated percentage change in the pedestrian-friendly environment index, with a BCR of 50% as the base scenario,[9] jumps from -50% to 0 when the BCR is reduced from 60% to 55%, that is, the highest slope (Table 4-2).

Based on the same analysis framework as above, the impacts of BCR on the other eco-city indexes are presented below (Tables 4-1 and 4-2). Firstly, when BCR falls from 45% to 40%, both of the green space accessibility indexes jump (the highest slope); this is because the collective open space of all four buildings in a block reaches the minimum level of a neighbourhood park for this study, i.e., 500 m^2. Secondly, both the trends of index values and accumulated percentages, as well as the negative values of elasticity, show that the lower the BCR, the more space for plants to grow, and hence the higher the adaptive capacity of CO_2 absorption and heat mitigation and improved wind environment. The highest slopes suggest that the maximum BCRs for these three indexes are 55%, 55%, and 30%, respectively. Finally, the daylight exposure index shows that the impact is not linear; the highest level occurs at BCRs of 65% and 55% and the lowest occurs at both the highest and lowest BCRs. This indicates that the tallest buildings with the lowest BCR of 30% and shortest buildings with the highest BCR of 75% both block the most sunlight, due to height and low permeability between buildings, respectively.

Overall, BCR is a moderately efficient planning tool for developing an eco-city; the lower the BCR, the more ecological the city. Furthermore, of all six eco-city aspects, BCR is the most efficient in the provision of green space for residents. The elasticity of overall eco-city index of -0.63 indicates that BCR is moderately (in)elastic in affecting the level of the eco-city built environment (Table 4-2). The trend of percentage changes of overall eco-city index shows that the ecological level increases as BCR decreases, and maximum BCRs of 55% and 40% are recommended for the ranges of BCR over and not greater than 40%, respectively. Of the six eco-city aspects, BCR is only elastic in affecting the accessibility to green space (Table 5-2) through the provision of a green community courtyard in the block. Though inelastic, lowering BCRs also significantly increases the levels of wind environment, pedestrian-friendly environment, open space for plants to grow, and green coverage rate for better CO_2 absorption and mitigation of urban heat.

---

residential areas in Taiwan, private property owners are encouraged to build sidewalk with a density bonus incentive.

[9] A BCR of 50% is set as the base scenario, from which the percentage change is calculated (refer to note 1 in Table 4-2). The percentage change statistics are applied to pinpoint the highest slope, where the most significant impact occurs.

3.2. Impacts of HDR

Both types of HDR are medium-efficient eco-city planning tools; the higher the HDR, the more ecological the city is, and the minimum recommended HDR is 1.5. HDR has a significant impact on green space accessibility. The elasticity values of HDR-S and HDR-B, representing the floor area added to the top of the building shifted from the side and back of the building, are 0.37 and 0.43, respectively (Table 5), less than the medium level of 0.5. The trend of the accumulated percentage change of overall eco-city index shows that the higher the HDR, and more ecological the city is. The most efficient point (i.e., the highest slope) occurs between HDR-S125 and HDR-B125 to 150; hence HDR 150 is recommended as the minimum level. This increase in open space arises because the collective backyards from the four building lots reach the minimum threshold at that point. Of the six eco-city aspects, only green space accessibility is elastic (i.e., greater than one). HDR is originally designed to ensure minimum lighting and solar radiation; however, based on the elasticity values, the increased solar radiation is achieved at a greater cost in terms of other aspects of ecological quality.

3.3. Impacts of SB

There are two types of SB, moving backwards (SB-B) and shifting floor area to the top of the building (SB-T), neither of which is a very efficient eco-city planning tool. However, the latter is slightly more efficient. A 3 m setback from the kerb is suggested as the minimum scenario to implement a greened sidewalk. Both types of SB are efficient in improving walkability, but SB-B will significantly lower green space accessibility if it downsizes the collective backyard to less than the minimum neighbourhood park size. The elasticity values of overall eco-city indexes of both types are positive but minimal, but SB-T is slightly more elastic, primarily because more open space is left on the ground (Table 6). The most efficient function of setback is providing more sidewalk space for pedestrians; therefore the accumulated percentage change in walkability jumps when the setback increases from 0 to 3 m. However, if SB-B continues to 4.5 m, the collective backyard open space becomes smaller than 500 m^2.

Table 4.1. Eco-city indexes, BCR

BCR	Eco-City Stakeholder/Aspect and Index									
	Stakeholder/Aspect	Residents				Plants	Natural Environment and Adaptive Capacity			Overall Eco-City Index (Mean)
		Pedestrian-Friendly Environment		Green Space Accessibility		Open Space for Plants	CO_2 Absorption, Heat Mitigation	Wind Environment	Daylight Exposure	
	Indexes	Sidewalk Percentage (%)	Green Sidewalk Percentage (%)	Green Open Space Percentage within 5-Minute Walk (%)	Distance to Nearest Green Open Space (m)	Green Coverage Ratio (%)	Green Coverage Ratio (%)	1-Ground Coverage Ratio (%)	Insolation (%)	
30%		100%	100%	18.0%	0 (1.00)[1]	86.6%	86.6%	79.2%	395,808 (49.7%)[2]	0.78
35%		100%	100%	16.9%	0 (1.00)	84.3%	84.3%	75.7%	423,215 (53.1%)	0.77
40%		100%	100%	15.7%	0 (1.00)	82.1%	82.1%	72.2%	432,619 (54.3%)	0.76
45%		100%	100%	7.3%	143 (0.00)	79.8%	79.8%	68.8%	430,112 (54.0%)	0.65
50% (Baseline)		100%	100%	7.3%	143 (0.00)	77.6%	77.6%	65.3%	425,599 (53.4%)	0.64

BCR	Eco-City Stakeholder/Aspect and Index								
Stakeholder/Aspect	Residents				Plants	Natural Environment and Adaptive Capacity			Overall Eco-City Index (Mean)
	Pedestrian-Friendly Environment		Green Space Accessibility		Open Space for Plants	CO_2 Absorption, Heat Mitigation	Wind Environment	Daylight Exposure	
Indexes	Sidewalk Percentage (%)	Green Sidewalk Percentage (%)	Green Open Space Percentage within 5-Minute Walk (%)	Distance to Nearest Green Open Space (m)	Green Coverage Ratio (%)	Green Coverage Ratio (%)	1-Ground Coverage Ratio (%)	Insolation (%)	
55%	100%	100%	7.3%	143 (0.00)	75.4%	75.4%	61.8%	449,875 (56.5%)	0.63
60%	100%	0.0%	7.3%	143 (0.00)	56.4%	56.4%	58.3%	404,469 (50.8%)	0.47
65%	100%	0.0%	7.3%	143 (0.00)	51.6%	51.6%	54.9%	454,768 (57.1%)	0.46
70%	100%	0.0%	7.3%	143 (0.00)	51.6%	51.6%	51.4%	425,711 (53.4%)	0.44
75%	100%	0.0%	7.3%	143 (0.00)	51.6%	51.6%	47.9%	393,897 (49.4%)	0.43
Mean	100%	60.0%	10.1%	100 (0.30)	69.7%	69.7%	63.5%	423,607 (53.2%)	0.60

1. The number in parentheses is the standardised index, obtained with the following equation: $= \left(1 - \frac{Distance\ to\ Nearest\ Park}{Longest\ Distance}\right) = \left(1 - \frac{Distance\ to\ Nearest\ Park}{143m}\right)$

2. The number in parentheses is the standardized index, obtained with the following equation $= \left(1 - \frac{\text{Insolation}}{\text{Insolation withpout Buildings}}\right) = \left(1 - \frac{\text{Solar Insolation}}{796.729(\text{Watts}/m^2)}\right)$

Table 4.2. Accumulated percentage change,[1] elasticity, and slope of eco-city index, BCR

BCR Scenario	Residents		Plants	Natural Environment and Adaptive Capacity			Overall
	Pedestrian-Friendly Environment	Green Space Accessibility	Open Space for Plants	CO_2 Absorption, Heat Mitigation	Wind Environment	Daylight Exposure	
30%	0%	1,524%	12%	12%	21%	-7%	22%
35%	0%	1,509%	9%	9%	16%	-1%	22%
40%	0%	1,493%	6%	6%	11%	2%	20%
45%	0%	0%	3%	3%	5%	1%	2%
50% (Baseline)	--	--	--	--	--	--	--
55%	0%	0%	-3%	-3%	-5%	6%	-1%
60%	-50%	0%	-27%	-27%	-11%	-5%	-30%
65%	-50%	0%	-34%	-34%	-16%	7%	- 31%
70%	-50%	0%	-34%	-34%	-21%	0%	- 33%
75%	-50%	0%	-34%	-34%	-26%	-7%	- 34%
Elasticity	-0.56	-16.93	-0.50	-0.50	-0.53	0.00 (0.25)[2]	-0.63
Highest Slope(s)	BCR60→55	BCR45→40	BCR60→55	BCR60→55	1.BCR45→40[3] 2.BCR60→55	BCR60→55	BCR60→55, BCR45→40

1. Percentage change = $\frac{(\text{Index}_{\text{target scenario}} - \text{Index}_{\text{base scenario}})}{\text{Index}_{\text{base scenario}}} \times 100\% = \frac{(\text{Index}_{\text{target scenario}} - \text{Index}_{\text{BCR50\%}})}{\text{Index}_{\text{BCR50\%}}} \times 100\%$

2. The impact of BCR on solar radiation is not linear; it peaks at BCRs of 55% and 65% and is lower at both ends. An alternative, gauging the elasticity between BCRs of 55% and 30%, shown in the parentheses, is provided.

3. For those with more than two highest slopes, the one with the highest index value is prioritised since it is more ecological.

Table 5. Accumulated percentage change, elasticity, and slope of eco-city indexes, HDR

HDR Scenario	Residents		Plants	Natural Environment and Adaptive Capacity			Overall
	Pedestrian-Friendly Environment	Green Space Accessibility	Open Space for Plants	CO_2 Absorption, Heat Mitigation	Wind Environment	Daylight Exposure	
HDR-S1.25	0%	-94%	-8%	-8%	-11%	8%	-21%
HDR-S150 (Baseline)	--	--	--	--	--	--	--
HDR-S1.75	0%	4%	5%	5%	8%	-1%	3%
HDR-S2.00	0%	6%	8%	8%	13%	-3%	4%
HDR-S2.25	0%	8%	11%	11%	18%	-5%	6%
Elasticity	0.00	1.54	0.29	0.29	0.43	-0.19	0.37
Highest Slope(s)	N/A	HDR-R1.25→1.50	HDR-R1.25 →1.50	HDR-R1.25 →1.50	HDR-S1.25 →1.50	HDR-R1.50→1.25	HDR-R1.25→1.50
HDR-B1.25	0%	-94%	-8%	-8%	-11%	-7%	-24%
HDR-B150 (Baseline)	--	--	--	--	--	--	--
HDR-B1.75	0%	4%	5%	5%	8%	2%	3%
HDR-B2.00	0%	6%	8%	8%	13%	-22%	1%
HDR-B2.25	0%	8%	11%	11%	18%	-21%	3%
Elasticity	0.00	1.53	0.29	0.29	0.43	-0.21	0.43
Highest Slope(s)	N/A	HDR-R1.25 →1.50	HDR-R1.25 →1.50	HDR-R1.25 →1.50	HDR-B1.25 →1.50	HDR-B1.75→2.00	HDR-R1.25→1.50

Table 6. Accumulated percentage change, elasticity, and slope of eco-city indexes, SB

SB Scenario	Residents		Plants	Natural Environment and Adaptive Capacity			Overall
	Pedestrian-Friendly Environment	Green Space Accessibility	Open Space for Plants	CO_2 Absorption, Heat Mitigation	Wind Environment	Daylight Exposure	
SB-B0	-100%	11%	-11%	-11%	0%	-6%	-31%
SB-B1.5	-50%	5%	-17%	-17%	0%	-1%	-20%
SB-B3 (Baseline)	--	--	--	--	--	--	--
SB-B4.5	0%	-94%	1%	1%	0%	-2%	-14%
SB-B6	0%	-94%	0%	0%	0%	-5%	-15%
Elasticity	0.50	-0.52	0.06	0.06	0.00	0.01	0.05
Highest Slope(s)	1. SB-B1.5→3 2. SB-B0→1.5	SB-B4.5→3	SB-B1.5→3	SB-B1.5→3	N/A	SB-B0→1.5	SB-B1.5→3
SB-T0	-100%	-3%	-18%	-18%	-14%	-7%	-34%
SB-T1.5	-50%	-1%	-19%	-19%	-0%	-3%	-22%
SB-T3 (Baseline)	--	--	--	--	--	--	--
SB-T4.5	0%	1%	4%	4%	6%	-3%	1%
SB-T6	0%	3%	8%	8%	12%	-2%	3%
Elasticity	0.50	0.03	0.13	0.13	0.13	0.02	0.19
Highest Slope(s)	SB-T1.5→3, SB-T0→1.5	SB-T 4.5→6	SB-T1.5→3	SB-T1.5→3	SB-T 0→1.5	SB-T0→1.5	SB-T1.5→3

3.4. Impacts of SW and BD

Both SW types—shifting floor area to the back (SW-B) and top (SW-T) of buildings—are inefficient eco-city planning tools, with elasticity values of -0.02 and 0.07, respectively (Table 7). The suggested minimum width is 9 m, because the highest slope occurs at this point where the size of the collective backyard exceeds the threshold of 500 m^2. The two types of BD—shifting floor to the side (BD-S) and to the top (BD-T) of buildings—affect the eco-city index very similarly to SW except that the distance between neighbouring buildings is shifted from the side to the back (Table 8).

3.5. Impacts of SCSB

The impacts of the two SCSB types—the horizontal (SCSB-H) and 4 m vertical (SCSB-V) setbacks—on eco-city development are minimal, but the former is slightly more efficient than the latter. For SCSB-H, the greater the street corner setback, the more ecological the city in terms of pedestrian-friendly environment, wind environment, open space for plants, CO_2 absorption, and heat mitigation. For SCSB-V, the setback at the first floor is more ecological than at the second floor and above. Both the elasticity values of overall eco-city index for both SCSB-H and SCSB–V are positive but barely above zero (Table 9), but the positive value for SCSB-H indicates that the higher the horizontal setback, the more ecological the city. This type of setback, specifically designed for more spacious street corners, has limited impact overall.

3.6. Impacts of Building Shape

Of the eight building shapes (tower (BS1), enclosing court (BS2), slab (BS3), U-shaped (BS4), convex-shaped (BS5), cross-shaped (BS6), X-shaped (BS7), and reverse U-shaped (BS8)), tower and X-shaped buildings are, on average, the most ecological, followed by the slab shape. The tower is superior because of its capacity to provide wide sidewalks for planting trees and higher insolation derived from its compact building shape; however, the X-shaped building is supreme due to its capacity to provide greened sidewalk and to concentrate backyards to form the equivalent of neighbourhood parks. The slab shape is relatively ecological in terms of green sidewalk and improved insolation through collective open space in front and back yards. Tower, X-shaped, and slab have the three highest overall eco-city index values of 0.64, 0.63, and 0.59, respectively (Table 10). The green sidewalk percentage index also reflects that these three building shapes, in addition to cross-shaped buildings, provide sidewalk wider than 3 m for growing plants. The X-shaped is the only building shape with collective backyards larger than 500 m^2, resulting in more green open space and zero distance to the nearest green open space. The tower and slab communities provide better insolation, with values of 53.4% and 54.4%, compared with 40% or below for other building shapes.

3.7. Efficiency of Planning and Design Tools for Eco-City Building

Based on the findings from the above simulation analysis of the hypothetical city, to improve the overall eco-city performance of the base scenario, reducing BCR is most efficient, though only moderately elastic, followed by increasing HDR-S and HDR-B, and increasing SB by shifting floor area to the top of the building (Table 11). Part of these findings are supported from the simulation analysis of the real case, such as BCR, SB, and HDR-B as the most efficient tools (-0.78, 0.50, and 0.41, respectively).

In certain cases, attention may be focused on improving certain individual aspects of an eco-city. The most effective tools for improving walkability are lowering BCR and implementing building setback (Table 11), also supported by the findings from the real case. Lowering BCR overwhelms other tools in increasing green-space accessibility, followed by a second-tier of increasing SW or BD by shifting floor area to the top of the building, and a third tier of increasing HDR. The most efficient way to increase open space for plants and street-level wind environment is to reduce BCR and increase HDR. However, HDB is found to be the most efficient way in these two regards for the real case. Finally, if insolation is an issue for a community, reducing HDR is the most appropriate choice in accordance with existing planning knowledge.

Table 7. Accumulated percentage change, elasticity, and slope of eco-city indexes, SW

SW Scenario	Residents		Plants	Natural Environment and Adaptive Capacity			Overall
	Pedestrian-Friendly Environment	Green Space Accessibility	Open Space for Plants	CO_2 Absorption, Heat Mitigation	Wind Environment	Daylight Exposure	
SW-B1.5 (Baseline)	--	--	--	--	--	--	--
SW-B2.5	0%	0%	1%	1%	0%	-4%	0%
SW-B3.5	0%	1%	1%	1%	0%	-6%	-1%
SW-B4.5	0%	1%	1%	1%	0%	-11%	-2%
SW-B5.5	0%	2%	0%	0%	0%	-17%	-3%
Elasticity	0.00	0.01	N/A	N/A	0.00	-0.06	-0.02
Highest Slope(s)	N/A	Limited impact	Limited impact	Limited impact	N/A	SW-B5.5→4.5	SW-B3.5→2.5
SW-T1.5 (Baseline)	--	--	--	--	--	--	--
SW-T2.5	0%	0%	3%	3%	3%	-3%	1%
SW-T3.5	0%	1%	4%	4%	6%	-10%	0%
SW-T4.5	0%	1493%	6%	6%	10%	-7%	18%
SW-T5.5	0%	1516%	7%	7%	13%	-9%	19%
Elasticity	0.00	5.68	0.03	0.03	0.05	N/A	0.07
Highest Slope(s)	N/A	SW-T3.5→4.5	SW-T1.5→2.5	SW-T1.5→2.5	SW-T3.5→4.5	SW-T3.5→2.5	SW-T3.5→4.5

Table 8. Accumulated percentage change, elasticity, and slope of eco-city indexes, BD

BD Scenario	Residents		Plants	Natural Environment and Adaptive Capacity			Overall
	Pedestrian-Friendly Environment	Green Space Accessibility	Open Space for Plants	CO_2 Absorption, Heat Mitigation	Wind Environment	Daylight Exposure	
BD-S1.5 (Baseline)	--	--	--	--	--	--	--
BD-S2.5	0%	0%	1%	1%	0%	6%	1%
BD-S3.5	0%	1%	1%	1%	0%	12%	2%
BD-S4.5	0%	1%	1%	1%	0%	15%	3%
BD-S5.5	0%	1%	0%	0%	0%	19%	3%
Elasticity	0.00	0.00	0.00	0.00	0.00	0.07	0.02
Highest Slope(s)	N/A	Limited impact	Limited impact	Limited impact	N/A	1. BD-S2.5→3.5, 2. BD-S1.5→2.5,	1. BD-S3.5→4.5, 2. BD-S2.5→3.5
BD-T1.5 (Baseline)	--	--	--	--	--	--	--
BD-T2.5	0%	0%	3%	3%	3%	5%	2%
BD-T3.5	0%	0%	4%	4%	6%	0%	2%
BD-T4.5	0%	1478%	6%	6%	10%	8%	21%
BDT-5.5	0%	1501%	7%	7%	13%	7%	22%
Elasticity	0.00	5.63	0.03	0.03	0.05	0.03	0.08
Highest Slope(s)	N/A	BD-T3.5→4.5	D-T1.5→2.5	BD-T1.5→2.5	BD-T3.5→4.5	BD-T4.5	BD-T3.5→4.5

Table 9. Accumulated percentage change, elasticity, and slope of eco-city indexes, SCSB

SCSB Scenario	Residents		Plants	Natural Environment and Adaptive Capacity			Overall
	Pedestrian-Friendly Environment	Green Space Accessibility	Open Space for Plants	CO_2 Absorption, Heat Mitigation	Wind Environment	Solar Radiation	
SCSB-H-1F0M (Baseline)	--	--	--	--	--	--	--
SCSB-H-1F2M	3%	0%	0%	0%	1%	6%	2%
SCSB-H-1F4M	7%	0%	1%	1%	3%	5%	3%
SCSB-H-1F6M	11%	0%	3%	3%	6%	3%	4%
Elasticity	0.04	0.00	0.01	0.01	0.02	N/A	0.02
Highest Slope(s)	SCSB-H-1F4M →1F6M	N/A	SCSB-H-1F4M →1F6M	SCSB-H-1F4M →1F6M	SCSB-H-1F4M □→1F6M	SCSB-H-1F2M	SCSB-H-1F0M →1F2M
SCSB-V-4M-None (Baseline)	--	--	--	--	--	--	--
SCSB-V-4M3F	0%	0%	0%	0%	0%	8%	1%
SCSB-V-4M2F	0%	0%	0%	0%	0%	8%	1%

Table 9. (Continued)

SCSB Scenario	Residents		Plants	Natural Environment and Adaptive Capacity			Overall
	Pedestrian-Friendly Environment	**Green Space Accessibility**	**Open Space for Plants**	**CO_2 Absorption, Heat Mitigation**	**Wind Environment**	**Solar Radiation**	
SCSB-V-4M1F	7%	0%	2%	2%	3%	5%	3%
Elasticity	0.02	0.00	0.01	0.01	0.01	N/A	0.01
Highest Slope(s)	SCSB-V-4M2F ·→4M1F	N/A	SCSB-V-4M2F ·→4M1F	SCSB-V-4M2F →4M1F	SCSB-V-4M2F ·→4M1F	SCSB-V-4M3F, -4M2F	SCSB-V-4M2F →4M1F

Table 10. Eco-city indexes, BS

BS	Stakeholder /Aspect	Residents				Plants	Natural Environment and Adaptive Capacity			Overall
		Pedestrian-Friendly Environment		Green Space Accessibility		Open Space for Plants	CO_2 Absorption, Heat Mitigation	Wind Environment	Daylight Exposure	
	Indexes	Sidewalk Percentage (%)	Green Sidewalk Percentage (%)	Green Open Space Percentage within 5-Minute Walk (%)	Distance to Nearest Green Open Space (m)	Green Coverage Rate (%)	Green Coverage Rate (%)	1-Ground Coverage Ratio	Insolation (%)	
BS1: Tower		100%	100%	7.3%	143 (0.00)	77.6%	77.6%	65.3%	425,599 (53.4%)	0.64
BS2-1.5:Enclosing Court		100%	0.0%	7.3%	143 (0.00)	61.2%	61.2%	65.3%	253,693 (31.8%)	0.46
BS2-2 (Baseline)		100%	0.0%	7.3%	143 (0.00)	58.8%	58.8%	65.3%	343,132 (43.1%)	0.47
BS2-2.5		100%	0.0%	7.3%	143 (0.00)	56.7%	56.7%	65.3%	293,552 (36.8%)	0.45
BS2-3		100%	100%	7.3%	143 (0.00)	72.7%	72.7%	65.3%	389,512 (48.9%)	0.61
Mean		100%	25.0%	7.3%	143 (0.00)	62.3%	62.3%	65.3%	319,972 (40.2%)	0.49

Table 10. (Continued)

BS	Stakeholder /Aspect	Residents				Plants	Natural Environment and Adaptive Capacity			Overall
		Pedestrian-Friendly Environment		Green Space Accessibility		Open Space for Plants	CO_2 Absorption, Heat Mitigation	Wind Environment	Daylight Exposure	
	Indexes	Sidewalk Percentage (%)	Green Sidewalk Percentage (%)	Green Open Space Percentage within 5-Minute Walk (%)	Distance to Nearest Green Open Space (m)	Green Coverage Rate (%)	Green Coverage Rate (%)	1-Ground Coverage Ratio	Insolation (%)	
BS3-1: Slab (Baseline)		100%	100%	7.3%	143 (0.00)	80.0%	80.0%	65.3%	421,792 (52.9%)	0.64
BS3-2		100%	0.0%	7.3%	143 (0.00)	76.5%	76.5%	65.3%	445,727 (55.9%)	0.55
Mean		100%	50.0%	7.3%	143 (0.00)	78.2%	78.2%	65.3%	433,760 (54.4%)	0.59
BS4-1: U-Shaped		100%	0.0%	7.3%	143 (0.00)	76.1%	76.1%	65.3%	406,114 (51.0%)	0.54
BS4-2 (Baseline)		100%	0.0%	7.3%	143 (0.00)	75.7%	75.7%	65.3%	287,033 (36.0%)	0.51
BS4-3		100%	0.0%	7.3%	143 (0.00)	75.3%	75.3%	65.3%	280,281 (35.2%)	0.51

BS	Stakeholder /Aspect	Residents				Plants	Natural Environment and Adaptive Capacity			Overall
		Pedestrian-Friendly Environment		Green Space Accessibility		Open Space for Plants	CO_2 Absorption, Heat Mitigation	Wind Environment	Daylight Exposure	
	Indexes	Sidewalk Percentage (%)	Green Sidewalk Percentage (%)	Green Open Space Percentage within 5-Minute Walk (%)	Distance to Nearest Green Open Space (m)	Green Coverage Rate (%)	Green Coverage Rate (%)	1-Ground Coverage Ratio	Insolation (%)	
Mean		100%	0.0%	7.3%	143 (0.00)	75.7%	75.7%	65.3%	324,476 (40.7%)	0.52
BS5-1: Convex-Shaped		100%	0.0%	7.3%	143 (0.00)	77.9%	77.9%	65.3%	300,123 (37.7%)	0.52
BS5-2 (Baseline)		100%	0.0%	7.3%	143 (0.00)	79.0%	79.0%	65.3%	294,317 (36.9%)	0.52
BS5-3		100%	0.0%	7.3%	143 (0.00)	80.2%	80.2%	65.3%	285,602 (35.8%)	0.53
Mean		100%	0.0%	7.3%	143 (0.00)	79.0%	79.0%	65.3%	293,348 (36.8%)	0.52
BS6-2: Cross-Shaped		100%	23.7%	7.3%	143 (0.00)	63.8%	63.8%	65.3%	309,999 (38.9%)	0.49
BS6-2.5 (Baseline)		100%	21.7%	7.3%	143 (0.00)	64.4%	64.4%	65.3%	319,580 (40.1%)	0.50

Table 10. (Continued)

BS	Stakeholder /Aspect	Residents				Plants	Natural Environment and Adaptive Capacity			Overall
		Pedestrian-Friendly Environment		Green Space Accessibility		Open Space for Plants	CO_2 Absorption, Heat Mitigation	Wind Environment	Daylight Exposure	
	Indexes	Sidewalk Percentage (%)	Green Sidewalk Percentage (%)	Green Open Space Percentage within 5-Minute Walk (%)	Distance to Nearest Green Open Space (m)	Green Coverage Rate (%)	Green Coverage Rate (%)	1-Ground Coverage Ratio	Insolation (%)	
BS6-3		100%	100%	7.3%	143 (0.00)	67.8%	67.8%	65.3%	416,884 (52.3%)	0.60
Mean		100%	48.5%	7.3%	143 (0.00)	65.3%	65.3%	65.3%	348,821 (43.8%)	0.53
BS7-2: X-Shaped		100%	24.6%	17.1%	0 (1.00)	66.3%	66.3%	65.3%	277,266 (34.8%)	0.59
BS7-25 (Baseline)		100%	26.1%	18.5%	0 (1.00)	68.0%	68.0%	65.3%	292,795 (36.7%)	0.60
BS7-3		100%	100%	20.1%	0 (1.00)	83.9%	83.9%	65.3%	312,421 (39.2%)	0.72
Mean		100%	50.2%	18.6%	0 (1.00)	72.7%	72.7%	65.3%	294,161 (36.9%)	0.63

BS	Stakeholder /Aspect	Residents				Plants	Natural Environment and Adaptive Capacity			Overall
		Pedestrian-Friendly Environment		Green Space Accessibility		Open Space for Plants	CO_2 Absorption, Heat Mitigation	Wind Environment	Daylight Exposure	
	Indexes	Sidewalk Percentage (%)	Green Sidewalk Percentage (%)	Green Open Space Percentage within 5-Minute Walk (%)	Distance to Nearest Green Open Space (m)	Green Coverage Rate (%)	Green Coverage Rate (%)	1-Ground Coverage Ratio	Insolation (%)	
BS8-1: Reverse U-Shaped		100%	0.0%	7.3%	143 (0.00)	76.1%	76.1%	65.3%	301,562 (37.9%)	0.51
BS8-2 (Baseline)		100%	0.0%	7.3%	143 (0.00)	75.7%	75.7%	65.3%	287,446 (36.1%)	0.51
BS8-3		100%	0.0%	7.3%	143 (0.00)	75.3%	75.3%	65.3%	272,552 (34.2%)	0.51
Mean		100%	0.0%	7.3%	143 (0.00)	75.7%	75.7%	65.3%	287,186 (36.0%)	0.51

3.8. Most Ecological Scenarios

Of the all of the scenarios considered, twelve have overall eco-city index values over 0.72 (Table 12). Of these twelve scenarios, only one is a non-tower building shape, i.e., X-shaped (BS 7), and the others are variations of tower-shaped buildings across five planning and design tools, consisting of large building setback with floor area shifted to the top (SB-T6, T4.5, and T3), smaller BCR (30%, 35%, and 40%), increasing side yard width by shifting floor area to top (SW-T11 and T9), increasing backyard depth by shifting floor area to top (BD-T11 and T9), and building setback by moving backward (SB-B3). All of these scenarios have streets with greened sidewalk the whole length of both sides, and a neighbourhood park equivalent is available in each block.

4. Impacts of Urban Planning and Design Tools on Animal-Biodiversity Eco-City Indexes

4.1. Impacts of BCR

Overall, BCR is an efficient tool for developing animal biodiversity in the built environment; the lower the BCR, the greater the animal biodiversity in the city. Furthermore, BCR is efficient in almost all the five indexes (Table 13). BCR levels of 40% and 55% are recommended for a positive impact on animal biodiversity.

4.2. Efficiency of Planning and Design Tools

The most efficient tools for improving overall animal biodiversity in the built environment are lowering BCR and increasing HDR-S and HDR-B (Table 14). In terms of increasing the number of patches, lowering BCR overwhelms other tools, followed by a second-tier of SW-T and BD-T and a third tier of increasing HDR-S or HDR-B. These tools have similar impacts on the land area of patches. Where the number of corridors is of primary concern, decreasing BCR is the most efficient tool, followed by increasing SB-B and SB-T. To raise the corridor quality in terms of fewer non-green intersections, the most efficient tool is reducing BCR.

Table 11. Elasticity of urban design tools

Planning Tools		Residents		Plants	Natural Environment (Adaptive)			Overall
		Pedestrian-Friendly Environment	**Green-Space Accessibility**	**Open Space for Plants**	**CO_2 Absorption, Heat Mitigation**	**Wind Environment**	**Daylight Exposure**	
1. BCR	1 BCR	**-0.56**[1]	**-16.93**	**-0.50**	**-0.50**	**-0.53**	0.00	**-0.63**
2. HDR	2-1 HDR-S: FA from Side	0.00	**1.54**	0.29	0.29	0.43	-0.19	0.37
	2-2 HDR-B: FA from Back	0.00	**1.53**	0.29	0.29	0.43	-0.21	0.43
3. SB	3-1 SB-B: Moving Backwards	**0.50**	0.03	0.06	0.06	0.00	0.01	0.05
	3-2 SB-T: FA to Top	**0.50**	**-0.52**	0.13	0.13	0.13	0.02	0.19
4. SW	4-1 SW-B: FA to Back	0.00	0.01	N/A	N/A	0.00	-0.06	-0.02
	4-2 SW-T: FA to Top	0.00	**5.68**	0.03	0.03	0.05	N/A	0.07
5. BD	5-1 BD-S: FA to Side	0.00	0.00	0.00	0.00	0.00	0.07	0.02
	5-2 BD-T: FA to Top	0.00	**5.63**	0.03	0.03	0.05	0.03	0.08
6. SCSB	6-1 SCSB-1F: Setbacks on 1F	0.04	0.00	0.01	0.01	0.02	N/A	0.02
	6-2 SCSB-4M: 4-Meter Setbacks	0.02	0.00	0.01	0.01	0.01	N/A	0.01

[1] Indexes with an elasticity of .5 or higher are shown in bold.

Table 12. Most ecological scenarios

Scenario	Residents				Plants	Natural Environment and Adaptive Capacity			Overall
	Pedestrian-Friendly Environment		Green Space Accessibility		Open Space for Plants	CO_2 Absorption, Heat Mitigation	Wind Environment	Daylight Exposure	
	Sidewalk Percentage	Green Sidewalk Percentage	Green Open Space Percentage within 5-Minute Walk	Distance to Nearest Green Open Space (M)	Green Coverage Rate	Green Coverage Rate (%)	1-Ground Coverage Ratio	Insolation (%)	
SB-T6	100%	100%	28.6%	0 (1.00)	93.0%	93.0%	81.3%	411,808 (51.7%)	0.79
SB-T4.5	100%	100%	26.9%	0 (1.00)	89.9%	89.9%	78.5%	408,193 (51.2%)	0.79
BCR-30	100%	100%	18.0%	0 (1.00)	86.6%	86.6%	79.2%	395,808 (49.7%)	0.78
BCR-35	100%	100%	16.9%	0 (1.00)	84.3%	84.3%	75.7%	423,215 (53.1%)	0.77
SB-T3	100%	100%	25.2%	0 (1.00)	86.4%	86.4%	75.7%	421,952 (53.0%)	0.77
BCR-40	100%	100%	15.7%	0 (1.00)	82.1%	82.1%	72.2%	432,619 (54.3%)	0.76
SW-T11	100%	100%	16.6%	0 (1.00)	82.5%	82.5%	69.6%	429,693 (53.9%)	0.75

Scenario	Residents				Plants	Natural Environment and Adaptive Capacity			Overall
	Pedestrian-Friendly Environment		Green Space Accessibility		Open Space for Plants	CO_2 Absorption, Heat Mitigation	Wind Environment	Daylight Exposure	
	Sidewalk Percentage	Green Sidewalk Percentage	Green Open Space Percentage within 5-Minute Walk	Distance to Nearest Green Open Space (M)	Green Coverage Rate	Green Coverage Rate (%)	1-Ground Coverage Ratio	Insolation (%)	
BD-T11	100%	100%	16.6%	0 (1.00)	82.5%	82.5%	69.5%	454,001 (57.0%)	0.75
SW-T9	100%	100%	14.9%	0 (1.00)	81.3%	81.3%	67.2%	439,889 (55.2%)	0.74
BD-T9	100%	100%	14.9%	0 (1.00)	81.3%	81.3%	67.3%	459,056 (57.6%)	0.74
SB-B3	100%	100%	15.8%	0 (1.00)	79.8%	79.8%	66.7%	416,946 (52.3%)	0.73
BS7-3: X-Shaped	100%	100%	20.1%	0 (1.00)	83.9%	83.9%	65.3%	312,421 (39.2%)	0.72

Table 13. Animal-biodiversity eco-city indexes, BCR

BCR	Patches		Corridor		Quality of Corridor	Overall
	Patches-Number1	Patches-Land Area Percentage2	T-Link-Number3	T-Link-Land Area4	1-(T-Gap)	
30%	**756 (100%)**	13%	64 (100%)	7%	85%	0.65
35%	756 (100%)	12%	64 (100%)	7%	85%	0.65
40%	756 (100%)	11%	64 (100%)	6%	85%	0.65
45%	12 (2%)	1%	64 (100%)	6%	85%	0.46
50%	12 (2%)	1%	64 (100%)	6%	85%	0.46
55%	12 (2%)	1%	64 (100%)	6%	85%	0.46
60%	12 (2%)	1%	8 (13%)	0%	6%	0.04
65%	12 (2%)	1%	8 (13%)	0%	5%	0.04
70%	12 (2%)	1%	8 (13%)	0%	5%	0.04
75%	12 (2%)	1%	8 (13%)	0%	5%	0.04
Elasticity	-68.89	-8.73	-0.97	-1.37	-1.04	-1.45
Highest Slope(s)	BCR45→40	BCR45→40	BCR60→55	BCR60→55	BCR60→55	BCR60→55, BCR45→40

1. The number in parentheses is a standardised index, obtained using the following equation: $= \left(1 - \frac{No.of\ Patches}{Largest\ No.of\ Patches}\right) = \left(1 - \frac{No.of\ Patches}{756}\right)$

2. The number is obtained using the following equation: $= \left(1 - \frac{Total\ Land\ Area\ of\ Patches}{Total\ Land\ Area}\right) = \left(1 - \frac{Total\ Land\ Area\ of\ Patches}{4{,}210{,}704m^2}\right)$

3. The number in parentheses is obtained using the following equation: $= \left(1 - \frac{No.of\ T{-}Links}{Largest\ No.of\ T{-}Links}\right) = \left(1 - \frac{No.of\ T{-}Links}{64}\right)$

4. The number is obtained using the following equation: $= \left(1 - \frac{Total\ Land\ Area\ of\ T{-}Links}{Total\ Land\ Area}\right) = \left(1 - \frac{Total\ Land\ Area\ of\ T{-}Links}{4{,}210{,}704M^2}\right)$

Table 14. Elasticity of animal-biodiversity eco-city indexes

Planning Tool	Patches		Corridor		Quality of Corridor	Overall
	Patches-Number	Patches-Land Area	T-Link-Number	T-Link-Land Area	1-(T-Gap)	
1 BCR	-68.89	-8.73	-0.97	-1.37	-1.04	-1.45
2-1 HDR-S	1.48	2.17	0.00	0.00	0.00	1.30
2-2 HDR-B	1.48	2.17	0.00	0.00	0.00	1.30
3-1 SB-B	-0.49	-0.89	0.44	0.69	0.47	0.18
3-2 SB-T	0.00	0.15	0.44	0.69	0.47	0.33
4-1 SW-B	0.00	0.00	0.00	0.00	0.00	0.00
4-2 SW-T	23.25	2.68	0.00	0.00	0.00	0.15
5-1 BD-S	0.00	0.00	0.00	0.00	0.00	0.00
5-2 BD-T	23.25	2.68	0.00	0.00	0.00	0.15
6-1 SCSB-1F	0.00	0.00	0.00	0.00	0.40	0.15
6-2 SCSB-4M	0.00	0.00	0.00	0.00	0.00	0.00

Table 15. Most animal-biodiversity built environment scenarios

Scenario	Patches		Corridor		Quality of Corridor	Overall
	Patches-Number	Patches-Land Area	T-Link-Number	T-Link-Land Area	1-(T-Gap)	
SB-T6	100%	21%	100%	8%	85%	0.66
SB-T4.5	100%	20%	100%	6%	85%	0.66
SB-T3	100%	18%	100%	5%	85%	0.66
BS7-3: X-Shaped	100%	15%	100%	6%	85%	0.65
BCR-30	100%	13%	100%	7%	85%	0.65
BCR-35	100%	12%	100%	7%	85%	0.65
BCR-40	100%	11%	100%	6%	85%	0.65
SW-T11	100%	12%	100%	6%	85%	0.65
BD-T11	100%	12%	100%	6%	85%	0.65
SW-T9	100%	10%	100%	6%	85%	0.64
BD-T9	100%	10%	100%	6%	85%	0.64
SB-B3	100%	11%	100%	5%	85%	0.64

4.3. Animal-Biodiversity Built Environment Scenarios

The twelve most efficient eco-city scenarios presented above are also the most efficient scenarios for animal biodiversity in the built environment, but in a slightly different order (Table 15). They vary only in land areas of patches and corridors, and the number of patches and corridors, as well as the quality of corridors, are the maximum afforded by the city layout.

Conclusion and Policy Implications

This research evaluates the efficiency of neighbourhood-level spatial planning and design tools for use in planning ordinances for eco-city building. Simulation analyses show that decreasing BCR is the most efficient of all the selected tools. It is efficient in almost all ecological aspects except for solar radiation, and the overall gain in all other aspects exceeds the loss in insolation. Increasing HDR is the second most efficient policy overall for eco-city building through provision of collective backyard open space between buildings, at the relatively small cost of lower insolation at the front of the building. SB-T is another relatively efficient tool in terms of providing more sidewalk space.

To enhance walkability, reducing BCR and setting back buildings are the two most efficient tools. Green space accessibility can be efficiently advanced by almost all tools except street-corner building setback, as long as the building mass is shifted to the tops of buildings. To provide more space to grow plants or turf and to improve the wind environment, and consequently the adaptive capacity of CO_2 absorption and heat mitigation, reducing BCR and increasing HDR are the two most efficient tools. Similar results were found for animal diversity in the built environment, which can be applied to the intended animal immigration corridors as necessary.

Tower-shaped buildings are more ecological than other building shapes. Generally, shifting space from the ground to the top of the building, rather than to the back or side, improves ecological quality. In other words, tall, skinny buildings create more ecological neighbourhoods than short, fat buildings if the ecological aspects of this study are adopted and assigned equal importance. It is worth noting that the findings of this research are limited to a hypothetical city, which differs from practical cases in terms of shape, size, orientation, and uniformity of buildings, as well as street layout. The actual impacts of planning and design tools on a specific neighbourhood of an eco-city rely on simulation analysis.

Acknowledgments

This research was funded by the Ministry of Science and Technology, Taiwan Government, NSC 99WFA0200051. The authors are grateful to the research assistance made by Yi-hsuan Chang and Ko-hsin Pai.

References

American Society of Civil Engineers (1999). *Wind Tunnel Studies of Buildings and Structures* (Manual of Practice No. 67). Virginia: ASCE.

Bertolini, L. F. le Clercq & Kapoen, L. (2005). Sustainable accessibility: a conceptual framework to integrate transport and land use plan-making. *Transport Policy*, 12(3), 207-220.

Brook McIlroy Planning + Urban Design (2006). *City of Burlington: Downtown Urban Design Guidelines* [online]. Available from: http://cms.burlington.ca/AssetFactory.aspx?did=4976 (Accessed 30 January 2012).

Camagni, R., Gibelli, M. C. & Rigamonti, P. (2002). Urban mobility and urban form: the social and environmental costs of different patterns of urban expansion. *Ecological Economics*, *40*(2), 199–216.

Cervero, R. (1996). Mixed land uses and commuting: Evidence from the American Housing Survey. *Transportation Research Part A: Policy and Practice*, *30*(5), 361–377.

Cervero, R., Ewing, R., Greenwald, M., Zhang, M., Walters, J., Feldman, M., Frank, L. & Thomas, J. (2011). Traffic generated by mixed-use developments—six-region study using consistent built environmental measures. *Journal of Urban Planning and Developmen*t, *137*(3), 248–261.

Cervero, R. & Sullivan, C. (2011). Green TODs: Marrying transit-oriented development and green urbanism. *International Journal of Sustainable Development & World Ecol*ogy, *18*(3), 210–218.

Costello, L. (2005). From prisons to penthouses: The changing images of high-rise living in Melbourne. *Housing Studies*, *20*(1), 49–62.

Devall, B. & Sessions, G. (1985). *Deep Ecology: Living As If Nature Mattered*. Salt Lake City: G. M. Smith.

Downton, P. F. (2009). *Ecopolis: Architecture and Cities for a Changing Climate*. Collingwood, Australia: CSIRO Publishing.

Environmental Systems Research Institute, Inc. (ESRI) (2007). *Calculating Insolation*. http://cms.burlington.ca/AssetFactory.aspx?did=4976.http://webhelp.esri.com/arcgisdesktop/9.2/index.cfm?TopicName=Calculating_solar_radiation.

Fisher-Gewirtzman, D., Pinsly, D. S., Wagner, I. A. & Burt, M. (2005). View-oriented three-dimensional visual analysis models for the urban environment. *Urban Design International*, *10*(1), 23–37.

Forman, R. T. T. & Godron, M. (1986). *Landscape Ecology*. New York: John Wiley & Sons.

Kasanko, M., Barredo, J. I., Lavalle, C., McCormick, N., Demicheli, L., Sagris, V. & Brezger, A. (2006). Are European cities becoming dispersed? A comparative analysis of fifteen European urban areas. *Landscape and Urban Planning*, *77*, 111–130.

Lo, R. H. (2009). Walkability: what is it? *Journal of Urbanism*, *2*(2), 145–166.

Martin, L. & March, L. (eds.) (1973). *Urban Space and Structures*. Cambridge: Cambridge University Press.

Ng, E., Chao, Y., Liang, C., Chao, R. & Jimmy, C. H. F. (2011). Improving the wind environment in high-density cities by understanding urban morphology and surface roughness: A study in Hong Kong. *Landscape and Urban Planning*, *101*, 59–74.

Oke, T. R. (1982). The energetic basis of the urban heat island. *Quarterly Journal of the Royal Meteorological Society*, *108*, 1–24.

Ramirez, L. K. B., Hoehner, C. M., Brownson, R. C., Cook, R., Orleans, C. T., Hollander, M., Barker, D. C., Bors, P., Ewing, R., Killingsworth, R., Petersmarck, K., Schmid, T. & Wilkinson, W. (2006). Indicators of activity-friendly communities: An evidence-based consensus process. *American Journal of Preventive Medicine*, *31*(6), 515–524.

Ratti, C. & Richens, P. (1999). Urban texture analysis with image processing techniques, in *Computers in Building: Proceedings of the CAADfutures'99 Conference*. USA: Kluwer Academic Publishers, 49–64.

Taipei City Government (2011). *Taipei City zoning ordinance [online, in Chinese]*. Available from: http://www.udd.taipei.gov.tw/pages/detail.aspx?Node=46&Page=2102&Index=5 (Accessed 2009).

Taipei City Government (2011). *Demographical Overview* [online]. Available from: http://english.taipei.gov.tw/ct.asp?xItem=1084529&ctNode=29491 &mp= 100002 (Accessed 18 August 2011).

Taiwan Ministry of The Interior. (2004). *Guidelines for Sidewalk Design in Urban Settings. Taipei: Minister of Interior[online, in Chinese]*. Available from: http://w3.cpami.gov.tw/district6/i5.htm (Accessed 20 January 2011).

Tsai, Y. H. (2010). Urban Planning/Design Tools for Improving Physical Aspects of Livability in High-Density Neighborhoods. 2011 International Winter Conference on Environmental Innovations and Sustainability. 28 and 29 January, 2011. Beppu, Oita, Japan.

U.S. Department of Justice. (1999). *Americans with Disabilities Act. Accessible Rights-of-Way: A Design Guide* [online]. Available from http://www.access-board.gov/prowac/guide/PROWGuide.htm (Accessed 20 January 2011).

Wong, M. S., Nichol, J. E., To, P. H. & Wang, J. (2010). A simple method for designation of urban ventilation corridors and its application to urban heat island analysis. *Building and Environment*, *45*, 1880–1889.

Yang, P. P. J., Putra, S. Y. & Li, W. J. (2007). Viewsphere: a GIS-based 3D visibility analysis for urban design evaluation. *Environment and Planning B: Planning and Design*, *43*, 971–992.

INDEX

C

F

G

H

I

J

K

L

M

N

O

P

S

T